Nathaniel L. Britton

A Preliminary Catalogue of the Flora of New Jersey

GEOLOGICAL SURVEY OF NEW JERSEY.

A PRELIMINARY

CATALOGUE OF THE FLORA

OF

NEW JERSEY.

COMPILED BY

N. L. BRITTON, PH. D.

WITH THE ASSISTANCE OF EMINENT BOTANISTS.

OFFICE OF THE SURVEY,
RUTGERS COLLEGE, NEW BRUNSWICK, N. J.

1881.

PRINTED BY
JOHN L. MURPHY, FINE BOOK PRINTER,
TRENTON, N. J.

*To His Excellency George C. Ludlow, Governor of the State of
New Jersey and President of the Board of Managers of the
State Geological Survey;*

DEAR SIR:—I have the honor to submit herewith a Preliminary
Catalogue of the Flora of New Jersey, compiled by N. L. Britton,
Ph. D., Botanist of the Survey.

It is designed to be used in perfecting the list of plants growing
in different parts of the State, by circulating copies of it among
botanists, so as to get their notes and corrections.

With high respect, your obedient servant,

GEO. H. COOK.

NEW BRUNSWICK, N. J., June 15, 1881.

———◆———

To Prof. Geo. H. Cook, State Geologist of New Jersey;

DEAR SIR:—I have the honor to hand you herewith a Preliminary Catalogue of the Flora of New Jersey.

Yours very respectfully,

N. L. BRITTON.

SCHOOL OF MINES,
COLUMBIA COLLEGE, N. Y., April 30, 1881.

(v)

PREFACE.

———◆———

This Catalogue of the Flora of New Jersey is intended for a preliminary tentative list of the plants hitherto found growing without cultivation in the State, with localities for all the rarer species. It is printed with interleaved pages, the blank ones being intended for additional notes on locality and distribution, and for corrections. A copy is sent to every amateur and professional botanist in New Jersey and the surrounding parts of other States as far as known.* It is requested that the books be returned to Prof. George H. Cook, State Geologist, New Brunswick, N. J., at the close of next season (1882), with all additions and corrections that botanists may report or suggest. It is hoped to obtain in this manner all that is at present known about the geographical and geological distribution, and frequency of occurrence of all the plants growing wild within the State. The final revised catalogue will be made up from these data at a future time.

LITERATURE OF THE SUBJECT.

The following authors have been consulted in the work of compilation :

John Torrey, M. D. "A Catalogue of Plants growing spontaneously within thirty miles of the city of New York." Albany, 1819. This is a one-hundred-paged pamphlet, containing many valuable notes on the occurrence of plants in the counties near New York.

---

* Additional copies may be obtained on application to the State Geologist.

P. D. Knieskern, M.D. "A Catalogue of Plants growing
without cultivation in the counties of Monmouth and Ocean,
N. J., being part of the Annual Report of Geological Survey of
New Jersey, for 1856." Forty-one pages, enumerating all the
Phanerogams and Acrogenous Cryptogams then known to grow
in those counties. Since that time many additions have been
made to this list.

Samuel Ashmead. "A List of Plants and a Catalogue of
Marine Algæ collected on the coast of Egg Harbor, at and near
Beesley's Point." Geological Report of Cape May county,
Trenton, 1857; pp. 149–154.

O. R. Willis, Ph.D. "Catalogue of Plants growing without
cultivation in the State of New Jersey." New York, 1874;
revised edition, 1878. This is a book of eighty-four pages and
a valuable contribution to our knowledge of the New Jersey
Flora.

C. F. Austin, in his sets of "Musci Appalachiani" and
"Hepaticæ Boreali-Americanæ," and in descriptions of many
new species of Mosses and Liverworts in the "Bulletin of the
Torrey Bontanical Club," and the "Botanical Gazette," con-
tributed largely to the literature of the Bryology of the State.

In 1870, William H. Leggett, editor of the "Bulletin of the
Torrey Club," began in the "Bulletin" the publication of a
revised catalogue of plants, with the same geographical range as
the one by Dr. Torrey, above mentioned. He was assisted by
the Club in this work, which is now finished up to the
Gramineæ, and has furnished a large number of valuable notes
on localities.

In addition to these longer lists, a number of notes on the
local Floras of various places have been from time to time published in the Torrey Club's "Bulletin," and have proved of
considerable use.

Mr. Addison Brown, Vice-President of the Torrey Club, has
made a study of the plants found growing on land made from
material brought in ballast, in the vicinity of New York, and
has published lists of them. (See Bull. Torr. Bot. Club, Vol.

VI., pp. 255 and 353, and Vol. VII., p. 122.) A list of those found by Mr. Brown and other botanists at Hoboken and Communipaw, together with a large number of additional species collected at Camden by Messrs. C. F. Parker, I. C. Martindale and Isaac Burk, and not described in Gray's "Manual of Botany," is inserted as an appendix to this catalogue. Those "ballast plants," which are recorded in Gray's "Manual," are admitted into the main catalogue. The plants are mostly natives of Europe, but some are from nearly every part of the globe. In time many species will be added to the list.

ARRANGEMENT OF THIS CATALOGUE.

In the arrangement of the catalogue, the sequence of succession of the Exogenous Orders is on the plan of Bentham and Hooker's "Genera Plantarum." The arrangement of the Endogenous Orders is that of Sir J. D. Hooker, in the English translation of Maout and Descaisne's "Botanique Descriptive et Analytique." The sequence of Genera under the Orders and of Species under the Genera is that of Gray's "Manual of Botany," except in the Liliaceæ where Watson's "Revision of the Liliaceæ" has been followed. The names are taken from Watson's "Index to American Botany," 1880, as far as that work is published, and chiefly from Gray's "Manual of Botany," 1880, for the remainder of the Phanerogams.

The arrangement of the Cryptogams is as follows:

Acrogens—Filices, Lycopodiaceæ, Equisetæ, Isoeteæ.
Anogens—Musci, Hepaticeæ.
Thallogens—Lichens, Fungi, Characeæ, Algæ.

The general geographical distribution of the phanerogamous Flora of New Jersey is of very great interest. The northern part of the State is covered with a soil composed of material brought from the north by the ice sheets of the glacial epoch, and consists of boulders and pebbles of many diffrent kinds of

rock, gravel, clay and sand, and this glacial drift soil extends as
far south as a line running irregularly from Perth Amboy, at
the mouth of the Raritan river, to Belvidere, on the Delaware
river. The region north of this line has a decidedly northern
Flora, over one-third of all the species growing wild in it
being natives of Europe, and a large number of the others being
only found further north on our own continent.

The southern part of the State is very unevenly covered with
a deposit of light-colored sand and gravel, with quartz pebbles,
whose origin is still uncertain. This is the "Yellow Drift,"*
frequently referred to in the following pages.

In this part of the State the northern Flora is meagre, and
twenty miles south of a line drawn from Perth Amboy to Trenton, is reduced to less than five per cent. of European species
with perhaps an equal number of northern North American
plants, and is replaced by an abundant truly American Flora
which is peculiar to this continent. These southern North
American plants are in a like manner but sparingly represented
on the glacial drift. The region included by these two diverg-
ing lines and the Delaware river, appears to possess a mixed
Flora, the northern species being found to some extent on the
mountainous portions of it, and the southern on less elevated
parts; which of these Floras is in excess, is yet to be determined,
but there are certainly many southern species there. Hence we
may in general conclude that the terminal glacial moraine is the
dividing line between the northern and southern Floras of New
Jersey.

In the work of compilation, my thanks are due to Mr. C. F.
Parker, of Camden, for exceedingly valuable assistance—he
kindly sent me catalogues of the New Jersey plants contained in
his herbarium, and compiled lists of the Mosses and Liverworts
from the collections of the late Coe F. Austin; to Mr. J. B.
Ellis, of Newfield, for the Catalogue of Fungi; to Rev. Francis
Wolle, of Bethlehem, Penna., and Rev. A. B. Hervey, of Taun-
ton, Mass., for lists of the Algæ; to Dr. T. F. Allen, of New

* See Annual Report of State Geologist for 1880, on the pre-glacial drift, pp. 87-97.

York, for the Characeæ; and to Prof. T. C. Porter, of Easton, Penna.; Prof. Samuel Lockwood, of Freehold; Dr. O. R. Willis, of White Plains, New York; Messrs. W. M. Wolfe and H. H. Rusby, of the North Jersey Botanical Club; Mr. W. H. Leggett, of New York; Mr. R. W. Brown, of Keyport, and Mr. Frank Tweedy, of Plainfield, for manuscript lists and notes on the Flora of different sections of the State.

N. L. BRITTON.

CONTENTS.

A PRELIMINARY

Catalogue of the Plants of New Jersey.

PHANEROGAMIA.

Class I.—EXOGENOUS PLANTS.

Sub-Class 1.—Angiospermæ.

Division A.—Polypetalæ.

RANUNCULACEÆ.

Clematis, L. . . . . Virgin's Bower. Clematis.

C. verticillaris, DC. Whorled-leaved Clematis. Rocky places in the middle and northern counties; rare. Preakness Mt., Passaic Co., W. L. Fischer; along the Delaware, above Phillipsburg, T. C. Porter, and near the Water Gap, W. M. Wolfe; Plainfield, sparingly, but in quantity three miles north of that place on First Newark Mt., Frank Tweedy.

C. Virginiana, L. Virgin's Bower. Travellers' Joy. Common in the northern and middle counties; rare on the Yellow Drift. Banks of Squan and Shark Rivers, P. D. Knieskern; near Holmdel, Monmouth Co., R. W. Brown.

Anemone, L. . . . . Wind-flower.

A. cylindrica, Gray. Cylindrical-fruited W. On dry limestone rocks about the zinc mines in Sussex Co., C. F. Austin,

A. Virginiana, L. Thimble-weed. Common in the northern and middle counties, but rare on the Yellow Drift. New Egypt, Ocean Co., and Middletown, Monmouth Co., P. D. Knieskern; near Chesquake, Middlesex Co., R. W. Brown.

A. dichotoma, L. Forked Anemone. (**A.** Pennsylvanica, L.) Banks of the Delaware, near Red Bank, Gloucester Co., C. F. Parker.

A. nemorosa, L. Wood Anemone. Throughout the State along the margins of woods and fencerows and in copses. Common.

(1)

A. Hepatica, L. (Hepatica triloba, L.) Liver-leaf. Round-lobed Hepatica. Common in the northern and northeastern parts of the State, but rare south of Monmouth Co. Eu.
A. acutiloba, Lawson. (Hepatica, L.) Sharp-lobed Hepatica. Montclair, Wm. Churchill. The only station reported in New Jersey.

Thalictrum, Tourn. . . . . MEADOW RUE.

T. anemonoides, Michx. Rue-leaved Anemone. Found throughout the State, growing in woods and along their margins, but is most abundant in the northern counties.
T. dioicum, L. Early Meadow Rue. Northern counties, generally in rocky places; not very common. New Brunswick, S. Lockwood; Princeton, O. R. Willis; frequent in Sussex Co., C. F. Austin; near Eagle Rock, Essex Co., Randall Spaulding.
T. purpurascens, L. Purplish M. Not common. Bergen Co., C. F. Austin.
Var. ceriferum, C. F. Austin. Waxy Meadow Rue. Rather common on the hillsides in the vicinity of New York and in the northern counties.
T. Cornuti, L. Common M. Low meadows and along streams. Common throughout.

Ranunculus, L. . . . . . CROWFOOT. BUTTERCUP.

R. aquatilis, L.; *Var.* stagnalis, DC. Stiff White Water Crowfoot. (R. divaricatus, Sch.) Squan and Shark Rivers, rare, P. D. Knieskern; Little Falls, W. M. Wolfe. Eu.
Var. trichophyllus, Chaix. Common White Water Crowfoot. Not unfrequent in slow streams. Cedar Brook, Plainfield, Frank Tweedy; Andover and Carpentersville, T. C. Porter. Eu.
R. multifidus, Pursh. Yellow Water Crowfoot. Rare. West of Hackensack, Saddle River, W. H. Leggett; Closter, C. F. Austin; Newton, Sussex Co., A. P. Garber; Swartzwood Lake, T. C. Porter.
R. ambiguus, Watson. Water-plantain Spearwort. (R. alismæfolius, Geyer.) Quite common in the northern parts of the State, and near New York City; rare in the southern counties.
R. Flammula, L.; *Var* reptens, Meyer. Creeping Spearwort. Along the Delaware above Phillipsburg, T. C. Porter. Eu.
R. pusillus, Poir. Weak Spearwort. Rare. Princeton, Dr. Torrey; Boonton, C. F. Austin; Camden, C. F. Parker; Verona, Essex Co., H. H. Rusby.
R. Cymbalaria, Pursh. Sea side Crowfoot. Head of Barnegat Bay, near Point Pleasant. rare, P. D. Knieskern.

R. abortivus, L. Small-flowered Crowfoot. Woods and along shaded streams. Common throughout.
Var. micranthus, Gray. Rare. Palisades, C. F. Austin; Montclair, Essex Co., W. M. Wolfe.
R. sceleratus, L. Cursed Crowfoot. Quite common near New York; New Egypt, rare, P. D. Knieskern; frequent about Lawrenceville, Mercer Co., O. R. Willis. Common in most localities. Eu.
R. recurvatus, Poir. Hooked Crowfoot. Rare south of Middlesex Co. Common in the northern counties, growing in woods.
R. Pennsylvanicus, L. Bristly Crowfoot. Rare near New York, and not found in the southern parts of the State. Princeton, Freehold and north, O. R. Willis; Camden, C. F. Parker; Newark Meadows, W. M. Wolfe. More common in the northern counties, but definite localities are desired.
R. fascicularis, Muhl. Early Crowfoot. Near Squan Village and Freehold, P. D. Knieskern; near Hightstown, O. R. Willis. Common in the northern and middle counties.
R. repens, L. Creeping Crowfoot. Buttercup. Grows in meadows throughout the State. Eu.
R. bulbosus, L. Bulbous Crowfoot. Buttercup. Found throughout the State, mostly near towns, but not very common anywhere except near New York. Nat. Eu.
R. acris, L. Tall Crowfoot. Buttercup. Common throughout, except in the pine barrens. Nat. Eu.

Caltha, L. . . . . . . MARSH MARIGOLD. COWSLIP.
C. palustris, L. Marsh Marigold. Common in the northern and middle counties, but not elsewhere. Camden, W. H. Redfield; sparingly in Mercer and Monmouth counties, O. R. Willis; along P. R. R., ten miles southwest of New Brunswick, Prof. Geo. H. Cook; near Keyport and Freehold, R. W. Brown. Eu.

Trollius, L. . . . . GLOBE FLOWER.
T. laxus, Salisb. Spreading Globe Flower. Rare. Confined to swamps in the northern parts of the State. Closter, C. F. Austin, A. Brown; Budd's Lake, T. C. Porter; near junction of Erie and N. R. R. of N. J., Jas. Hyatt; abundant near Passaic, G. C. Woolson; Sussex Co., A. P. Garber.

Coptis, Salisb. . . . . GOLDTHREAD.
C. trifolia, Salisb. Three-leaved Goldthread. Sparingly in the western part of Mercer Co., O. R. Willis; Trenton, W. M. Wolfe; about Budd's Lake, T. C. Porter; New Durham, W. H. Leggett; Preakness,

W. L. Fischer; Closter, C. F. Austin; Sussex Co., A. P. Garber; Succasunny, T. C. Porter. Not found in the southern counties. Eu.

Helleborus, L. . . . . HELLEBORE.

H. viridis, L. Green Hellebore. Warren Co., F. Knighton; near Freehold, S. Lockwood; West Orange, Essex Co., W. M. Wolfe. Adv. Eu.

Aquilegia, Tourn. . . . . COLUMBINE.

A. Canadensis, L. Wild Columbine. Near Keyport and Freehold, R. W. Brown; hills back of Princeton, O. R. Willis; Squan, Monmouth Co., P. D. Knieskern; shady sand hills, Atlantic City, C. F. Parker, and common on rocks in the northern counties.

A. vulgaris, L. Belvidere, F. Knighton. Adv. Eu.

Delphinium, Tourn. . . . . LARKSPUR.

D. Consolida, L. Field Larkspur. Fields, Plainfield, Frank Tweedy; ballast ground at Communipaw, A. Brown; and at Camden, C. F. Parker; Closter, C. F. Austin; Long Branch, C. F. Parker. Nat. Eu.

Hydrastis, L. . . . . ORANGE-ROOT.

H. Canadensis, L. Orange-root. Warren Co., F. Knighton. Rare. Austin's specimens were collected near Port Jervis, N. Y., and not in Sussex Co., (C. F. Parker).

Actæa, L. . . . . BANEBERRY.

A. spicata, L.; *Var.* rubra, Michx. Red Baneberry. Princeton and Lawrenceville, Mercer Co., and Cream Ridge, Monmouth Co., O. R. Willis; Franklin, Essex Co., W. M. Wolfe; Preakness Mt., W. L. Fischer; near Keyport, R. W. Brown; Plainfield, F. Tweedy; and not uncommon in the northern counties. Rare or absent south of Monmouth Co. Eu.

A. alba, Bigel. White Baneberry. Rather common in the northern and middle counties, but rare elsewhere in the State.

Cimicifuga, L. . . . . BUGBANE.

C. racemosa, Nutt. Black Snake-root. Lawrenceville, Mercer Co., Lanning; and frequent in the northern counties and near New York; Chesquake Creek, Middlesex Co., R. W. Brown; Somerset Gap, Frank Tweedy; New Brunswick, S. Lockwood.

MAGNOLIACEÆ.
Magnolia, L. . . . . MAGNOLIA.

M. glauca, L. Laurel Magnolia, Swamp Sassafras, Sweet Bay. New Durham, C. F. Austin; Short Hills, near Plainfield, W. M. Wolfe;

South Amboy, N. L. Britton; and common in swamps on cretaceous and tertiary soil in the southern counties, occurring only very rarely north of the terminal glacial moraine.

Liriodendron, L. . . . . TULIP-TREE.

L. Tulipifera, L. Tulip-tree. Whitewood. Common in the northern counties and as far south as the Raritan River, and Mercer Co.; frequent in Burlington and Monmouth counties; less common further south. Varies with the wood from white to yellow in color, and also in toughness.

ANONACEÆ.

Asimina, Adans. . . . . NORTH AMERICAN PAPAW.

A. triloba, Dunal. Common Papaw. Bridgeton, I. C. Martindale.

MENISPERMACEÆ.

Menispermum, L. . . . . MOONSEED.

M. Canadense, L. Canadian Moonseed. Frequent in the northern and middle counties. Monmouth Co., Dr. Torrey; near Holmdel, Monmouth Co., R. W. Brown.

BERBERIDACEÆ.

Berberis, L. . . . . BARBERRY.

B. vulgaris, L. Common Barberry. Near Matteawan and Sandy Hook, Monmouth Co., R. W. Brown; Bergen Hills, W. H. Leggett; Red Bank, P. D. Knieskern; Closter, C. F. Austin. Not common. Nat. Eu.

Caulophyllum, Michx. . . . . . BLUE COHOSH.

C. thalictroides, Michx. Pappoose-root. Pascack, C. F. Austin; Plainfield, Frank Tweedy; Preakness, W. L. Fischer; Milburn, Essex Co., H. H. Rusby. Rare.

Podophyllum, L . . . MAY APPLE. MANDRAKE.

P. peltatum, L. May Apple. Common in the northern and middle parts of the State. Cream Ridge, Monmouth Co., and Princeton, O. R. Willis; near Perth Amboy, C. A. Hollick; near Holmdel, Monmouth Co., R. W. Brown. Rare further south.

NYMPHACEÆ.

Brasenia, Schreber. . . . . WATER-SHIELD.

B. peltata, Pursh. Water-shield. Barrsville, Ocean Co., and Shark River, Monmouth Co., P. D. Knieskern; Collier's Mills, Ocean Co., N.

L. Britton, and frequent in ponds in the pine barrens; Swartzwood Lake, Lake Hopatcong, Sussex Co., H. H. Rusby; Camden, C. F. Parker; Passaic River above Paterson, H. H. Rusby.

Nelumbium, Juss. . . . NELUMBO. SACRED BEAN.

N. luteum, Willd. Yellow Nelumbo. Pond near Woodstown, Salem Co., a long-known locality; Swartzwood Lake, Sussex Co., T. C. Porter, H. H. Rusby.

Nymphæa, Tourn. . . . . WATER LILY.

N. odorata, Ait. Odorous White Water Lily. Ponds and slow streams; common throughout.

Var. minor, Sims. Small White Water Lily. Near Atco, I. H. Hall; and elsewhere in the pine barrens; in some places more common than the type; Budd's Lake, Morris Co., C. F. Parker.

Nuphar, Smith. . YELLOW POND LILY. SPATTER DOCK.

N. advena, Ait. Common Yellow Pond Lily. Ponds and ditches. Common throughout.

N. pumilum, Smith. Small Yellow Pond Lily. Common in the Hackensack River, etc., near Closter, C. F. Austin. The only station reported in the State. Eu.

SARRACENIACEÆ.

Sarracenia, Tourn. . . . SIDE SADDLE FLOWER.

S. purpurea, L. Pitcher Plant. Huntsman's Cup. Common in cedar swamps in the pine barrens, and in peat bogs all over the State.

PAPAVERACEÆ.

Papaver, L. . . . . POPPY.

P. dubium, L. Smooth-fruited Corn Poppy. Ballast ground at Communipaw, Frank Tweedy; cultivated ground between Camden and White Horse, C. F. Parker. Adv. Eu.

P. somniferum, L. Common Poppy. Escaped into waste soil at Plainfield, Frank Tweedy. Adv. Eu.

Argemone, L. . . . . PRICKLY POPPY.

A. Mexicana, L. Mexican Prickly Poppy. Waste places, Ocean and Monmouth Cos., not common, P. D. Knieskern; ballast and waste grounds, Camden, C. F. Parker. Adv. Mexico.

Chelidonium, L. . . . . CELANDINE.
C. majus, L. Celandine. Waste grounds, Ocean and Monmouth Cos., P. D. Knieskern; Camden, C. F. Parker, and frequent near New York. Adv. Eu.

Sanguinaria, Dill. . . . . BLOOD-ROOT.
S. Canadensis, L. Canadian Blood-root. Common in the northern and middle counties; rare in the pine barrens and southern parts of the State. New Egypt, Ocean Co., very rare, P. D. Knieskern; near Keyport, R. W. Brown.

Glaucium, Tourn. . . . . HORN POPPY.
G. luteum, Scop. Yellow Horn Poppy. Princeton, rare, O. R. Willis; in ballast at Communipaw, Addison Brown. Adv. Eu.

FUMARIACEÆ.

Adlumia, Raf. . . . . . . CLIMBING FUMITORY.
A. cirrhosa, Raf. Climbing Fumitory. Palisades, C. F. Austin; Belvidere, F. Knighton; near Greenwood Lake, W. M. Wolfe. Rare.

Dicentra, Bork. . . . . . DUTCHMAN'S BREECHES.
D. Cucullaria, DC. Dutchman's Breeches. Rather common on rocks in the northern and middle counties, but very rare south of the red sandstone. Keyport, R. W. Brown; three miles northwest of New Brunswick, Prof. Geo. H. Cook.
D. Canadensis, DC. Squirrel Corn. Mountains of Sussex Co., C. F. Austin.
D. eximia, DC. Delaware Water Gap, C. F. Austin.
(These latter two plants are not in Austin's collection; C. F. Parker.)

Corydalis, Vent. . . . . CORYDALIS.
C. glauca, Pursh. Pale Corydalis. Frequent on rocks in the northern counties; not found south of the trias. Palisades, C. F. Austin; Bloomsbury, A. P. Garber; near Phillipsburg, T. C. Porter; First Mt., Essex Co., W. M. Wolfe.
C. aurea, Willd. Golden Corydalis. Princeton, Mercer Co., O. R. Willis, on the authority of Dr. Torrey. Very rare.
C. flavula, Raf. Yellowish Corydalis. Banks of the Delaware, Camden, very rare, C. F. Parker; below Holland Station, Hunterdon Co., T. C. Porter; Cape May Co., C. F. Austin.

Fumaria, L. . . . . Fumitory.

F. officinalis, L. Common Fumitory. Princeton and Hightstown, Mercer Co., O. R. Willis; in ballast, Communipaw, Addison Brown; and Camden, C. F. Parker. Adv. Eu.

CRUCIFERÆ.

Nasturtium, R. Br. . . . . Water Cress.

N. officinale, R. Br. True Water Cress. Camden, rare, C. F. Parker; occasional in streams in the vicinity of New York; Plainfield, F. Tweedy; Bloomsbury, T. C. Porter; Hanover, Morris Co., Great Swamp, Morris Co., and at New Brunswick, Prof. Geo. H. Cook. Nat. Eu.

N. sylvestre, R. Br. Yellow Cress. Banks of the Delaware near the Waterworks, and in ballast at Camden, C. F. Parker; Bloomfield, Essex Co., H. H. Rusby. Nat. Eu.

N. palustre, DC.; *Var.* hispidum, Fisch. and Meyer. Marsh Cress. Hackensack Meadows, C. F. Austin; Camden, C. F. Parker; near Bloomfield, Essex Co., W. M. Wolfe; near Phillipsburg, T. C. Porter; Weehawken, N. L. Britton; apparently not very common. Eu.

N. lacustre, Gray. Lake Cress. Swartzwood Lake, T. C. Porter. The only station known in this part of the country.

N. Armoracia, Fries. Horse Radish. Escaped from gardens into wet places along brooks and ditches in many places. Nat. Eu.

Dentaria, L. . . . . Toothwort. Pepper-root.

D. diphylla, L Two-leaved Pepper-root. Norwood, Bergen Co., O. R. Willis; Tappan, and common in Sussex Co., C. F. Austin. Rare and confined to the northern counties.

D. laciniata, Muhl. Cut-leaved Pepper-root. Frequent in the northern and middle counties. Weehawken, C. F. Austin; Freehold, O. R. Willis; Camden, W. M. Canby; Hoboken Hills, W. H. Leggett; Verona, Essex Co., H. H. Rusby.

Cardamine, L. . . . . , ... Bitter Cress.

C. rhomboidea, DC. Spring Cress. Frequent throughout the State. Closter, C. F. Austin; Camden, C. F. Parker; rare in Monmouth and Ocean Cos., P. D. Knieskern. Common near New York.

C. rotundifolia, Michx. Mountain Water Cress. Cool shaded springs, Middletown, Monmouth Co., very rare, P. D. Knieskern.

C. pratensis, L. Cuckoo Flower. Cedar swamp at New Durham, C. F. Austin, W. H. Leggett. Rare.

C. hirsuta, L. Small Bitter Cress. Wet places. Common in the northern and middle counties; rare on the Yellow Drift.

Var. sylvatica, Gray. Small Bitter Cress. On rocks in the northern counties. Palisades and Hoboken, W. H. Leggett.

Arabis, L. . . . . ROCK CRESS.

A. lyrata, L. Lyrate-leaved Rock Cress. Sparingly on rocks in the northern counties. Little Falls, W. M. Wolfe ; First Mt., Essex Co., H. H. Rusby; near Holmdel, Monmouth Co., R. W. Brown; near Budd's Lake, T. C. Porter; shady places, Ocean and Monmouth Cos., not common, P. D. Knieskern. Occasionally grows in sand.

A. hirsuta, Scop. Hairy Rock Cress. Mostly confined to rocky places in the northern counties, and not common. Sussex Co., C. F. Austin, A. P. Garber; near Hightstown, Mercer Co., O. R. Willis.

A. lævigata, DC. Smooth Rock Cress. Rocky places middle and northern parts of the State. Not very common. First Newark Mt., one mile north of Plainfield, Frank Tweedy; and Essex Co., H. H. Rusby; common on the Palisades, C. F. Austin.

A. Canadensis, L. Sickle-pod. Common in the northern and middle counties ; rare on the Yellow Drift.

Barbarea, R. Br. . . . . WINTER CRESS.

B. vulgaris, R. Br. Common Winter Cress. Yellow Rocket. Common in fields and along roadsides, except in the pine barrens. Nat. Eu.

B. præcox, R. Br. Early Winter Cress. In ballast, Camden, C. F. Parker; Communipaw, Addison Brown; Newark Neck, W. M. Wolfe. Adv. Eu.

Erysimum, L. . . . . TREACLE MUSTARD.

E. cheiranthoides, L. Worm-seed Mustard. Banks of the Hackensack, C. F. Austin, perhaps native there, but also in ballast at Communipaw, Addison Brown, where it is adventive from Europe.

Sisymbrium, L. . . . . . . HEDGE MUSTARD.

S. officinale, Scop. Hedge Mustard. Common along roadsides and near dwellings throughout the State. Nat. Eu.

S. Thaliana, Gand. Mouse-ear Cress. Sparingly in fields. Near Evona, Union Co., Frank Tweedy ; Sandy ground, near Bergen Point, W. M. Wolfe; Belvidere, F. Knighton ; Atlantic Co., J. H. Redfield ; and near New York. Nat. Eu.

S. canescens, Nutt. Tansy Mustard. Shore of Delaware Bay, Cape May Co., C. F. Austin. The only locality reported.

S. Sophia, L. In ballast, Camden, C. F. Parker, and Communipaw, Addison Brown. Adv. Eu.

Brassica, Tourn. . . . . . . MUSTARD, TURNIP.

B. Sinapistrum, Boiss. Charlock. Rather common in cultivated fields throughout the State. Nat. Eu.

B. alba, L. White Mustard. Ballast and waste ground, Camden, C. F. Parker; Communipaw, N. L. Britton; Ocean and Monmouth Cos., P. D. Knieskern. Not common. Adv. Eu.

B. nigra, L. Black Mustard. Common throughout in fields and waste places. Nat. Eu.

B. campestris, L. Turnip. In cultivated fields, and in ballast at Communipaw, C. F. Parker. Adv. Eu.

Draba, L. . . . . WHITLOW GRASS.

D. Caroliniana, Walt. Carolina Whitlow Grass. Sparingly in sandy fields in the southern counties. Camden, W. M. Canby ; Burlington, Burlington Co., Isaac Burk.

D. verna, L. Whitlow Grass. Quite common in sandy fields and along roadsides throughout the State, and probably in part introduced and naturalized from Europe. Eu.

Alyssum, Tourn. . . . . ALYSSUM.

A. maritimum, L. Sweet Alyssum. In ballast at Camden, C. F. Parker. Adv. Eu.

A. calycinum, L. In ballast at Communipaw, A. Brown. Adv. Eu.

Camelina, Crantz. . . . . FALSE FLAX.

C. sativa, Crantz. False Flax. Fields and waste grounds, frequent. In ballast at Communipaw, A. Brown. Adv. Eu.

Capsella, Vent. . . . . . . SHEPHERD'S PURSE.

C. Bursa-pastoris, Mœnch. Shepherd's Purse. Very common in cultivated fields and waste places throughout. Nat. Eu.

Thlaspi, Tourn, ... PENNYCRESS.

T. arvense, L. Mithridate Mustard. Ballast, Communipaw, Addison Brown ; and Camden, C. F. Parker. Adv. Eu.

Lepidium, L. . . . . PEPPERGRASS.

L. Virginicum, L. Wild Peppergrass. Common throughout as a weed along roadsides, etc.

L. ruderale, L. Wild Peppergrass. Frequent near New York, and in ballast at Camden, I. C. Martindale. Adv. Eu.

L. intermedium, Gray. Wild Peppergrass. Ballast at Camden, I. C. Martindale. Adventive from northwestern States.

L. campestre, L. Field Peppergrass. Waste and cultivated grounds; becoming very common near New York. Plainfield, Frank Tweedy; Camden, W. M. Canby; Freehold, S. Lockwood. Nat. Eu.

L. Draba. In ballast at Communipaw, Addison Brown. Adv. Eu.

Senebiera, DC. . . WART CRESS. SWINE CRESS.

S. didyma, Pers. Wart Cress. Ballast at Camden, C. F. Parker; and at Communipaw, Addison Brown. Adventive from the South.

S. Coronopus, DC. Wart Cress. With the last species. Adv. Eu.

Cakile, Tourn. . . . . SEA ROCKET.

C. Americana, Nutt. American Sea Rocket. Common on the sea-beach along the whole coast.

C. maritima, Scop. Sea Rocket. In ballast at Communipaw, A. Brown; and Camden, C. F. Parker. Adv. South.

Raphanus, L. . . . . RADISH.

R. Raphanistrum, L. Wild Radish. Jointed Charlock. Rape. A troublesome weed in cultivated fields throughout the State. Nat. Eu.

CAPPARIDACEÆ.

Polanisia, Raf. . . . . POLANISIA.

P. graveolens, Raf. Clammy-weed. Long Branch, I. H. Hall; Bergen, P. V. LeRoy.

RESEDACEÆ.

Reseda, L. Mignonette. . . DYER'S ROCKET.

R. Luteola, L. Dyer's Weed. In ballast at Camden, C. F. Parker; and Communipaw, A. Brown. Adv. Eu.

VIOLACEÆ.

Ionidium, Vent. (Solea, Ging.) . . GREEN VIOLET.

I. concolor, Benth. and Hook. One-colored Green Violet. On Bool's Island, Delaware River, I. C. Moyer; near Milford, Hunterdon Co., T. C. Porter. Rare.

Viola, L. . . . . . VIOLET. HEART'S EASE.

V. rotundifolia, Michx. Round-leaved Violet. Confined to the northern counties and grows only sparingly there. Morristown, W. H. Leggett; Closter, Bergen Co., C. F. Austin; Warren Co., T. C. Porter; Verona, Essex Co., H. H. Rusby; woods back of Tenafly, W. H. Leggett; Stanhope, C. F. Austin.

V. lanceolata, L. Lance-leaved Violet. Quite common through-out the State.

V. primulæfolia, L. Primrose-leaved Violet. Frequent throughout the State.

V. blanda, Willd. Sweet White Violet. Common throughout.

V. cucullata, Ait. Common Blue Violet. Common throughout.

Var. palmata, Gray. Hand-leaved Violet. Rather common everywhere in damp ground.

Var. cordata, Gray. Heart-leaved Violet. Franklin, Essex Co., H. H. Rusby; Preakness Mt., W. L. Fischer; Plainfield, F. Tweedy.

V. sagittata, Ait. Arrow-leaved Violet. Common throughout.

V. pedata, L. Bird-foot Violet. Common in sandy or gravelly soil, particularly in the southern and middle counties. Sometimes found with pink or even white flowers.

Var. bicolor, Gray. Pansy Violet. Marble Hill, near Phillipsburg, Warren Co., T. C. Porter. Rare and local.

V. canina, L.; *Var.* sylvestris, Regel. Dog Violet. Sparingly through the middle and northern parts of the State. Freehold, O. R. Willis; Bergen Co., C. F. Austin; frequent near New York. Eu.

V. rostrata, Muhl. Long-spurred Violet. Sparingly in the northern counties. Marble Hill, near Phillipsburg, T. C. Porter; Little Falls, W. M. Wolfe; Bergen Co., C. F. Austin; Hemlock Falls, South Orange, W. H. Leggett; near Watchung Station, N. Y. and G. L. R. R., W. M. Wolfe; Plainfield, Frank Tweedy.

V. striata, Ait. Pale Violet. Rare. Closter, Bergen Co., C. F. Austin; Newark, W. H. Rudkin; Verona, Essex Co., H. H. Rusby.

V. Canadensis, L. Canada Violet. Northern part of the State, F. Knighton; Palisades, Bergen Co., C. F. Austin. Very rare.

V. pubescens, Ait. Downy Yellow Violet. Quite common throughout.

Var. eriocarpa, Nutt. Cream Ridge, Monmouth Co., O. R. Willis; Verona, Essex Co., H. H. Rusby.

Var. scabriuscula, Torr. and Gray. Bergen Co., C. F. Austin; Franklin, Essex Co., H. H. Rusby. Rare.

V. tricolor, L. Pansy, Heart's Ease. In ballast, Communipaw, Addison Brown; fields near Trenton, O. R. Willis. Nat. Eu.

Var. arvensis, Ging. Near New Egypt, Ocean Co., N. L. Britton; Nat. Eu. (?). In ballast at Camden, C. F. Parker. Adv. Eu.

CISTACEÆ.

Helianthemum, Tourn. . . . . . . ROCK-ROSE.

H. Canadense, Michx. Frost-weed. Common throughout in dry, sandy, or gravelly soil.

H. corymbosum, Michx. Frost-weed. Sandy soil in the pine barrens near the coast, but rare. Ocean and Monmouth Counties, P. D. Kneiskern.

Hudsonia, L. . .   . . HUDSONIA.

H. ericoides, L Heath-like Hudsonia. Common throughout the southern parts of the State. Not found north of the Yellow Drift.

H. tomentosa, Nutt. Wooly Hudsonia. Common in the sands of the seashore along the whole coast, and sparingly a few miles inland in the pine barrens. Quaker Bridge, Burlington Co., C. F. Parker.

Lechea, L. . . . . PINWEED.

L. major, Michx. Large Pinweed. Common in dry sandy places throughout.

L. thymifolia, Michx. (L. Novæ-Cæsareæ, Aust.) Rather common all over the State. Closter, C. F. Austin; Long Branch and the pine barrens, W. H. Leggett; Tom's River and Camden, C. F. Parker.

L. minor, Lam. Small Pinweed. Common throughout the State, except in the pine barrens, where it seems to be mostly replaced by the

Var. (?) pulchella, Leggett. Beautiful Pinweed. Pleasant Mills, W. H. Leggett; Manchester, N. L. Britton, M. Ruger; and probably throughout the southern and eastern counties; Atsion, W. M. Canby; Quaker Bridge, J. S. Merriam, W. H. Leggett.

L. racemulosa, Michx. Racemed Pinweed. The prevailing and most common form of Lechea in the pine barrens and probably sparingly in other parts of the State. It has been included under the protean L. minor, of Dr. Gray's Manual, from which it must be distinguished. See Bull. Torr. Bot. Club, Vol. VI., 251.

L. maritima, Leggett (L. thymifolia, Pursh). Abundant in the sands of the sea-shore along the whole coast, and in the sands of the pine barrens.

L. tenuifolia, Michx. Small-leaved Pinweed. Phalanx, Monmouth Co., W. H. Leggett.

POLYGALACEÆ.

Polygala, Tourn.   . .   . . . MILKWORT.

P. lutea, L. Yellow Milkwort. Common in damp sandy ground in the pine barrens and confined to the Yellow Drift.

P. incarnata, L. Pink Milkwort. Rare. In sandy ground, Camden Co., C. F. Parker; Haddonfield, W. M. Canby.

P. sanguinea, L. Red Milkwort. In damp sandy soil. Common throughout.

P. Nuttallii, Torr. & Gray. Nuttall's Milkwort. Southern and middle counties. Rare. Long Branch, T. F. Allen; Pine Barrens, W. M. Canby.

P. cruciata, L. Cross Milkwort. Common along the margins of swamps in the pine barrens, and sparingly in the middle counties. South Amboy, T. F. Allen.

P. brevifolia, Nutt. Short-leaved Milkwort. With the last species, and probably nearly as common in the southern counties. Secaucus Swamp, T. F. Allen.

P. verticillata, L. Whorled Milkwort. Dry sandy soil, common throughout the State.

P. ambigua, Nutt. Doubtful Milkwort. Common in Ocean and Monmouth counties, P. D. Knieskern; Franklin, Essex Co., H. H. Rusby. Generally grows with P. verticillata, Nutt., and should properly be considered as a variety of that plant.

P. fastigiata, Nutt. Red Milkwort. Pine barrens of Ocean Co., not common, P. D. Knieskern; pine barrens of New Jersey, Nuttall in Gray's Manual.

P. senega, L. Seneca Snake-root. "In open woods and on hills, N. J."—Torrey Catalogue, 1819. No definite localities are reported for this plant, but it should be found within our limits.

P. paucifolia, Willd. Fringed Polygala Rare, and confined to the northern and middle counties. New Durham Swamp, Torrey Catalogue, but not collected there recently; Franklin, Essex Co., H. H. Rusby; near Freehold, O. R. Willis; Morris Co., C. F. Austin; Warren Co., C. F. Parker.

P. polygama, Walt. Polygamous Milkwort. Old fields, Closter, C. F. Austin; Deal, Monmouth Co., Geo. Smith; Sea Bright, Monmouth Co., N. L. Britton; Gloucester Co., C. F. Parker. Not common.

CARYOPHYLLACEÆ.

Dianthus, L. . . . . CARNATION. PINK.

D. Armeria, L. Deptford Pink. Sandy fields and roadsides. Rather common throughout. Nat. Eu.

D. prolifer, L. Proliferous Pink. Roadside near Haddonfield, Camden Co., C. F. Parker. Adv. Eu.

Saponaria, L. . . . . SOAPWORT.

S. officinalis, L. Bouncing Bet. Roadsides and waste places. Common throughout. Nat. Eu.

Vaccaria, Medik. . . . . COW-HERB.

V. vulgaris, Host. Common Cow-herb. In ballast at Camden, C. F. Parker; and Communipaw, Addison Brown. Also occasional along

roadsides in other parts of the State. Monmouth and Ocean Cos., P.
D. Knieskern; near Passaic, G. C. Woolson. Adv. Eu.

Silene, L. . . . . . CATCHFLY. CAMPION. PINK.

S. stellata, Ait. Starry Campion. Found throughout the State, but
most common in the middle counties.

S. inflata, Smith. Bladder Campion. Closter. C. F. Austin; New
Durham, M. Ruger; Montclair, H. H. Rusby; Little Falls, Passaic
Co., W. M. Wolfe; and in ballast at Communipaw, Addison Brown;
and Camden, C. F. Parker. Nat. Eu.

S. Pennsylvanica, Michx. Wild Pink. Sparingly throughout the
State.

S. Virginica, L. Fire Pink. Near Camden, W. M. Canby; Warren
Co., F. Knighton. Rare.

S. Armeria, L. Sweet William Catchfly. Mercer Co., Dr. John
Torrey; Bergen, Ocean and Cape May Cos., C. F. Parker; Union Co.,
Frank Tweedy; near Red Bank and Keyport, S. Lockwood. Not
common. Adv. Eu.

S. antirrhina, L. Sleepy Catchfly. Closter, C. F. Austin; waste
places and ballast at Camden, C. F. Parker; Verona, Essex Co., H. H.
Rusby. Frequent near New York.

S. noctiflora, L. Night-flowering Catchfly. Warren Co., F. Knighton; ballast at Camden, C. F. Parker. Nat. Eu.

Lychnis, Tourn. . . . . . . COCKLE. LYCHNIS.

L. vespertina, Sibth. Evening Lychnis. In ballast at Camden, C.
F. Parker; and Communipaw, A. Brown; Newark Meadows along C.
R. R. of N. J., W. M. Wolfe. Adv. Eu.

L. Githago, Lam. Corn Cockle. Frequent in wheat fields. In
ballast at Camden, C. F. Parker. Adv. Eu.

Arenaria, L. . . . . SANDWORT.

A. serpyllifolia, L. Thyme-leaved Sandwort. Roadsides and waste
places. Common throughout. Nat. Eu.

A. squarrosa, Michx. Pine Barren Sandwort. Common in the
southern counties, growing in pure sand. Most abundant in the pine
barrens, and confined to the area of the Yellow Drift.

A. stricta, Michx. Michaux's Sandwort. Rare and confined to
rocky places in the northern counties. Cooper's Furnace, Phillipsburg, A. P. Garber; Hunterdon Co., T. C. Porter.

A. lateriflora, L. Sparingly throughout the State. Sussex Co., A.
P. Garber; Budd's Lake, Morris Co., T. C. Porter; Closter, C. F.
Austin; Atlantic City, C. F. Parker; Plainfield, F. Tweedy. Eu.

A. peploides, L. Sea-side Sandwort. Frequent in sands of the sea-shore. Eu.

Stellaria, L. . . . . . . CHICKWEED. STARWORT.

S. media, Smith. Common Chickweed. Cultivated and waste grounds ; very common in all parts of the State. Nat. Eu.

S. longifolia, Muhl. Long-leaved Starwort. Common in damp meadows in the northern and middle counties; rare on the Yellow Drift. Eu.

Cerastium, L. . . . . . MOUSE-EAR CHICKWEED.

C. vulgatum, L. Mouse-ear Chickweed. Rare. Bergen Co., C. F. Austin; shady rocks along N. R. R. above junction with Erie R. R., T. F. Allen. Nat. Eu.

C. viscosum, L. Mouse-ear Chickweed. Grassy fields and copses. Common throughout. Nat. Eu.

C. nutans, Raf. Sticky Chickweed. Sparingly in damp rocky places. Bergen Co., C. F. Austin; Bloomfield, Essex Co., H. H. Rusby ; Palisades near Tenafly, N. L. Britton ; Ocean and Monmouth Cos., common (?), P. D. Knieskern.

C. oblongifolium, Torrey. Oblong-leaved Chickweed. Palisades, C. F. Austin ; P. V. Le Roy. Very rare.

C. arvense, L. Field Chickweed. Warren Co., F. Knighton ; Fort Lee, T. F. Allen ; on the Palisades near Tenafly, N. L. Britton ; banks of the Delaware near Camden, C. F. Parker. Mostly confined to the northern parts of the State, and not common. Eu.

Sagina, L. . . . . PEARLWORT.

S. procumbens, L. Creeping Pearlwort. Damp places; not common. Sidewalks of Camden, C. F. Parker. Eu.

S. apetala, L. Non-petaled Pearlwort. Sandy places. Rare or more probably overlooked. Mercer and Monmouth Cos., Dr. Torrey in Willis' Catalogue. Eu.

S. decumbens, Torr. & Gray. (**S.** subulata, Wimmer.) Creeping Pearlwort. In ballast at Camden, C. F. Parker. Eu.

Var. Smithii, Gray. Smith's Pearlwort. Somers Point, Atlantic Co., C. E. Smith.

Lepigonum, Fries. (Spergularia, L.) SAND SPURREY,

L. medium, Fries. Sand Spurrey. Common in salt marshes. Eu.

L. rubrum, Fries. Sand Spurrey. Frequent in sandy soil along the coasts. Eu.

L. salinum, Fries. Sand Spurrey. In salt marshes, but much rarer than L. medium, at least about New York. Ballast at Camden, C. F. Parker, and at Communipaw, W. M. Wolfe. Eu.

Spergula, L. . . . . Spurrey.
S. arvensis, L. Corn Spurrey. Occasional in cultivated fields and in ballast. Adv. Eu.

PARONYCHIÆ.

Anychia, Michx. . . . . . . Forked Chickweed.
A. dichotoma, Michx. Forked Chickweed. Common throughout.

Scleranthus, L. . . . . Knawel.
S. annuus, L. Annual Knawel. Common throughout. Nat. Eu.

PORTULACACEÆ.

Portulaca, Tourn. . . . . Purslane.
P. oleracea, L. Common Purslane. Cultivated and waste grounds. Common throughout. Nat. Eu.

Claytonia, L. . . . . Spring Beauty.
C. Virginica, L. Virginian Spring Beauty. Common in the middle and northern counties. Banks of the Delaware, Camden, C. F. Parker; near Keyport, R. W. Brown; near New Egypt, Ocean Co., P. D. Knieskern.

ELATINACEÆ.

Elatine, L. . . . . Water-wort.
E. Americana, Arnott. American Water-Wort. Not common. Banks of the Passaic, T. F. Allen; tidal mud, Delaware river at Camden, C. F. Parker; Lake Hopatcong, Morris Co., T. C. Porter.

HYPERICACEÆ.

Ascyrum, L. . . . . . . St. Peter's-wort.
A. stans, Michx. Erect St. Peter's-wort. Damp places in the southern counties; abundant in the pine barrens, and confined to the area of the Yellow Drift.
A. Crux-andreæ, L. St. Andrew's Cross. Southern counties, with the same general distribution as the last species. Also at Bergen Point, W. H. Leggett.

Hypericum, L. . . . . Sᴛ. Jᴏʜɴ's-ᴡᴏʀᴛ.

H. pyramidatum, Ait. Great St. John's-wort. On the Delaware below Phillipsburg, T. C. Porter; near Phillipsburg, A. P. Garber. Rare, and confined to the north-western part of the State.

H. prolificum, L. Shrubby St. John's-wort. Wet pine barrens, Manchester, Ocean Co., P. D. Knieskern; near Shark River Station, N. J. S. R. R., R. W. Brown. Not nearly so common as the next species.

H. densiflorum, Pursh. Shrubby St. John's-wort. Frequent throughout the pine barren country, and confined to the Yellow Drift.

H. adpressum, Bart. St. John's-wort. Rare. Closter, C. F. Austin; Freehold, O. R. Willis; Tenafly, Joseph Schrenck.

H. angulosum, Michx. Angled St. John's-wort. Swamps in the pine barrens. In the cedar swamp, at Weehawken, Torrey Catalogue, 1819.

H. ellipticum, Hook. St. John's-wort. In a sphagnous swamp, near Camden, E. Diffenbaugh.

H. perforatum, L. Common St. John's-wort. Fields and meadows. Common throughout. Nat. Eu.

H. corymbosum, Muhl. Dotted St. John's-wort. Damp places. Frequent throughout the State.

H. mutilum, L. Low St. John's-wort. Wet grounds. Common throughout.

H. Canadense, L. Canadian St. John's-wort. Wet sandy places. Common throughout.

Var. major, Gray. Camden, C. F. Parker.

H. Sarothra, Michx. Orange Grass. Pine-weed. Sandy fields and roadsides. Common throughout.

Elodes, Adans. . . . . . Mᴀʀꜱʜ Sᴛ. Jᴏʜɴ's-ᴡᴏʀᴛ.

E. Virginica, Nutt. Common in swamps throughout the State.

MALVACEÆ.

Althæa, L. . . . . Mᴀʀꜱʜ Mᴀʟʟᴏᴡ.

A. officinalis, L. Common Marsh Mallow. Salt marshes, but not very common. Nat. Eu.

Malva, L. . . . . Mᴀʟʟᴏᴡ.

M. rotundifolia, L. Common Mallow. Waste places. Common. Nat. Eu.

M. sylvestris, L. High Mallow. Sparingly along roadsides, and in ballast at Camden. Adv. Eu.

M. moschata, L. Musk Mallow. Sussex Co., C. F. Austin ; Plainfield, Frank Tweedy. Adv. Eu.

Sida, L. . . . . Sida.

S. spinosa, L. Spiny Sida. Frequent in waste places. New Egypt, Ocean Co., P. D. Kneiskern ; Camden, C. F. Parker; Bridgeton, N. L. Britton; ballast at Communipaw, A. Brown. Nat. Tropical America.

Abutilon, Tourn. . . . . Indian Mallow.

A. Avicennæ, Gærtn. Velvet-leaf. Waste places. Rather common throughout.

Kosteletzkya, Presl. . . . . . . Kosteletzkya.

K. Virginica, Gray. Virginian Kosteletzka. Salt marshes on the coast. Rather scarce. Ocean and Monmouth Cos., P. D. Knieskern ; Cape May, W. M. Canby ; Hackensack Meadows, T. F. Allen.

Hibiscus, L. . . . . Rose Mallow.

H. Moscheutos, L. Swamp Rose Mallow. Along salt marshes and brackish ditches. Rather common. Bound Brook, Frank Tweedy!

H. Trionum, L. Bladder Ketmia. Cultivated fields and about gardens. Ballast at Camden, C. F. Parker. Adv. Eu.

TILIACEÆ.

Tilia, L. . . . . Linden. Basswood.

T. Americana, L. Basswood. Rather common in the northern counties, but rare elsewhere. Banks of Squan River, P. D. Knieskern; Cliffwood, near Keyport, S. Lockwood.

Var. pubescens, Loud. Basswood. On the high hills near Weehawken, Torrey Catalogue, 1819. Not since noted.

LINACEÆ.

Linum, L. . . . . Flax.

L. Virginianum, L. Common Wild Flax. Dry sandy woods. Common throughout.

L. striatum, Walt. Winged Flax. Low swampy ground. Quite common.

L. sulcatum, Riddell. Grooved Flax. Rare. Palisades, Bergen Co., and in Sussex Co., C. F. Austin.

L. usitatissimum, L. Common Flax. Ballast and waste ground, Camden, C. F. Parker; ballast at Communipaw, A. Brown. Adv. Eu.

GERANIACEÆ.

Geranium, L.   . . . . GERANIUM. CRANESBILL.

G. maculatum, L. Wild Cranesbill. Open woods. Common throughout.

G. Carolinianum, L. Carolina Cranesbill. Barren and waste places. Quite common.

G. dissectum, L. Cut-leaved Cranesbill. Ballast grounds at Camden, C. F. Parker; Little Falls, Passaic Co., W. M. Wolfe; Milburn, Essex Co., H. H. Rusby. Adv. Eu.

G. columbinum, L. Long-stalked Cranesbill. Chatham, Morris Co., W. H. Leggett. Nat. Eu.

G. pusillum, L. Small-flowered Cranesbill. Waste places and ballast at Camden, C. F. Parker; in ballast at Communipaw, A. Brown; Keyport, R. W. Brown, Plainfield, F. Tweedy. Nat. Eu.

G. Robertianum, L. Herb Robert. Common in rocky places in the northern and middle counties; rare in the southern parts of the State. Atlantic City, C. F. Parker. Eu.

Erodium, L'Her.   . . . . STORKSBILL.

E. cicutarium, L'Her. Storksbill. Near Franklin, Essex Co., H. H. Rusby; Woodbury, W. M. Canby; and ballast at Camden, C. F. Parker; and Communipaw, M. Ruger; College Farm, New Brunswick, Prof. Geo. H. Cook. Adv. Eu.

Flœrkea, Willd.   . . . . FALSE MERMAID.

F. proserpinacoides, Willd. False Mermaid. Rare. Pascack and Closter, C. F. Austin; Franklin, Essex Co., H. H. Rusby.

Impatiens, L.   . . . .   BALSAM. JEWEL-WEED.

I. pallida, Nutt. Pale Touch-me-not. Frequent in the northern counties; rare in other parts of the State. Snake Hill, Newark Meadows, T. F. Allen; Weehawken, W. H. Leggett; base of Palisades opposite Riverdale, N. Y., E. P. Bicknell.

I. fulva, Nutt. Spotted Touch-me-not. Wet places. Common throughout the State. A form with white flowers was reported from near Tom's River, Ocean Co., by Dr. P. D. Knieskern.

Oxalis, L.   . . . . WOOD SORREL.

O. violacea, L. Violet Wood Sorrel. Frequent in rocky woods. Most abundant in the middle counties.

O. stricta, L. Yellow Wood Sorrel. Fields, copses and roadsides. Common throughout.

RUTACEÆ.

Xanthoxylum, Colden. . . . . . . PRICKLY ASH.

X. Americanum, Mill. Toothache-tree. Closter, C. F. Austin; near Freehold, O. R. Willis; Verona, Essex Co., H. H. Rusby. Rare.

SIMARUBACEÆ.

Ailanthus, Desf. . . . . AILANTHUS.

A. glandulosus, Desf. Tree of Heaven. Becoming widely naturalized, and is found nearly all over the State. Adv. China.

ILICINEÆ.

Ilex, L. . . . . HOLLY.

I. opaca, Ait. American Holly. Abundant in the southern and eastern counties as far north as Sandy Hook. Most abundant on the area covered by the Yellow Drift.

I. verticillata, Gray. Black Alder, Winterberry. Swampy ground. Rather common throughout the State.

I. lævigata, Gray. Smooth Winterberry. Low ground near Camden, C. F. Parker; pine barrens, W. M. Canby; Tom's River, N. L. Britton; near Keyport, R. W. Brown; Secaucus Swamp and Chatham, W. H. Leggett. Common only in the southern counties.

I. glabra, Gray. Inkberry. Frequent in the pine barrens, and generally in the southern counties. New Durham and Secaucus Swamps, W. H. Leggett.

Nemopanthes, Raf. . . . . MOUNTAIN HOLLY.

N. Canadensis, DC. Mountain Holly. Sussex Co., A. P. Garber; Camden, C. F. Parker; Hackensack Swamps, W. H. Leggett; Ocean Co., rare, P. D. Knieskern; Budd's Lake, Morris Co., T. C. Porter; pine barrens, W. M. Canby. Not common.

CELASTRACEÆ.

Celastrus, L. . STAFF-TREE. SHRUBBY BITTER-SWEET.

C. scandens, L. Wax-work. Climbing Bitter-sweet. Thickets and along streams. Most common in the middle and northern counties.

Euonymus, Tourn. . . . . . . SPINDLE-TREE.

E. atropurpureus, Jacq. Burning-bush. Rare. Near Paterson on the road to Hamburg, W. L. Fischer; near Little Falls, H. H. Rusby.

E. Americanus, L. Strawberry Bush. Freehold, O. R. Willis;
Closter, Bergen Co., C. F. Austin; near Camden, C. F. Parker; near
Milburn, Essex Co., N. L. Britton; Keyport, R. W. Brown; Freehold,
S. Lockwood.

Var. obovatus, Torr. and Gray. Probably as abundant as the type,
Franklin, Essex Co., H. H. Rusby.

RHAMNACEÆ.

Rhamnus, Tourn. . . . . BUCKTHORN.

R. cathartica, L. Common Buckthorn. Near Haddonfield, C. F.
Parker. Nat. Eu.

R. alnifolia, L'Her. Alder-leaved Buckthorn. Ditches by side of
the railroad near New Durham, W. H. Leggett, T. F. Allen.

R. Caroliniana, Walt. (Frangula, L.) Carolina Buckthorn. Secaucus Swamp, W. H. Leggett; near New Durham Station, 1879, N. L.
Britton.

Ceanothus, L. . . . , . . . NEW JERSEY TEA.

C. Americanus, L. New Jersey Tea. Open woods. Common
throughout.

VITACEÆ.

Vitis, Tourn. . . . . GRAPE.

V. Labrusca, L. Northern Fox Grape. Moist Thickets. Common
throughout.

V. æstivalis, Michx. Summer Grape. Thickets. Common throughout.

V. cordifolia, Michx. Frost Grape. Not so common as the pre-
ceding species. Camden, C. F. Parker; Ocean Co., P. D. Knieskern;
Summit, W. H. Leggett, (V. riparia, Michx.?); common in Essex
Co., H. H. Rusby; Keyport, R. W. Brown.

Ampelopsis, Michx. . . . . AMERICAN IVY.

A. quinquefolia, Michx. Virginian Creeper. Woods and along
streams. Rather common throughout the State.

SAPINDACEÆ.

Staphylea, L. . . . . BLADDER-NUT.

S. trifolia, L. American Bladder-nut. Palisades, W. H. Leggett;
banks of Squan River, rare, P. D. Knieskern; Plainfield, F. Tweedy;
on a bank just west of Paterson, H. H. Rusby; Little Falls, W. M.
Wolfe; Snake Hill, M. Ruger; New Brunswick, Geo. H. Cook.

Acer, Tourn. . . . . MAPLE.

A. Pennsylvanicum, L. Striped Maple. Sparingly in the northern counties. Sussex Co., C. F. Austin; high hills of New Jersey, Torrey Catalogue, 1819.

A. spicatum, Lam. Mountain Maple. Warren Co., J. H. Redfield, F. Knighton; on First Mt., Essex Co., W. M. Wolfe.

A. saccharinum, Wang. Sugar Maple. Frequently found native in the northern counties, and very extensively planted throughout.

A. dasycarpum, Ehrh. Silver Maple. It is uncertain whether this tree is a native of the State or not. I have never seen it growing where it could not be traced to cultivation. It is very commonly used as a shade and ornamental tree.

A. rubrum, L. Red Maple, Swamp Maple. Swamps and wet grounds. Common in all parts of the State.

Negundo, Mœnch. . . . . BOX ELDER.

N. aceroides, Mœnch. Ash-leaved Maple. Rare. Near Paterson, and Red Bank, Monmouth Co., W. H. Leggett; banks of Hackensack River, half a mile west of Closter, C. F. Austin; Green Brook, Union Co., Frank Tweedy.

ANACARDIACEÆ.

Rhus, L. . . . . SUMACH.

R. typhina, L. Staghorn Sumach. Rocky hillsides; confined to the northern parts of the State. Banks of the Delaware, Warren Co., C. F. Parker; Closter, C. F. Austin; Essex Co., along Orange Mt., H. H. Rusby.

R. glabra, L. Smooth Sumach. Common in the northern and middle counties, and sparingly on the Yellow Drift.

R. copallina, L. Dwarf Sumach. Rather common throughout the State in poor soil.

R. venenata, DC. Poison Sumach. Poison Dogwood. Swamps. Common throughout.

R. Toxicodendron, L. Poison Ivy. Poison Oak. Thickets and fence-rows. Common throughout.

Var. radicans, Torrey. Frequent in swampy places.

LEGUMINOSÆ.

Lupinus, Tourn. . . . . LUPINE.

L. perennis, L. Wild Lupine. Sandy soil. Quite common throughout the State.

Crotallaria, L. . . . . RATTLE-BOX.

C. sagittalis, L. Rattle-box. Sandy soil. Quite common throughout the State.

Trifolium, L. . . . . . CLOVER. TREFOIL.

T. arvense, L. Rabbit-foot Clover. Barren sandy fields. Common throughout. Nat. Eu.

T. pratense, L. Red Clover. Fields and pastures. Common throughout. Nat. Eu.

T. repens, L. White Clover. Common in all parts of the State. Nat. Eu.

T. agrarium, L. Yellow or Hop Clover. Sparingly in fields throughout the State. Ballast at Camden, C. F. Parker. Nat. Eu.

T. procumbens, L. Low Hop Clover. Not so common as the last species. Franklin, Essex Co., H. H. Rusby; Freehold, O. R. Willis; in ballast at Communipaw, A. Brown. Nat. Eu.

Var. minus, Koch. Occasionally found with the type. Nat. Eu.

T. incarnatum, L. Shamong, W. M. Canby. Adv. Eu.

Melilotus, Tourn. . . . . SWEET CLOVER.

M. officinalis, Willd. Yellow Melilot. Sparingly in waste places. Camden, C. F. Parker; Holmdel, Monmouth Co., R. W. Brown; Hightstown and Freehold, O. R. Willis; Communipaw, T. F. Allen, A. Brown; Bloomfield, Essex Co., W. M. Wolfe. Adv. Eu.

M. alba, Lam. White Melilot. Waste places. Frequent. Adv. Eu.

Medicago, L. . . . . MEDICK.

M. sativa, L. Lucerne Clover. Stockton township, Camden Co., I. C. Martindale; ballast at Camden, C. F. Parker. Occasionally cultivated for fodder. Adv. Eu.

M. lupulina, L. Black Medick. Nonesuch. Waste places. Common throughout. Adv. Eu.

M. maculata, Willd. Spotted Medick. Waste places and ballast. Not common. Franklin, Essex Co., H. H. Rusby; Camden, C. F. Parker; Communipaw, A. Brown. Adv. Eu.

M. denticulata, Willd. Toothed Medick. Waste places and ballast at Camden, C. F. Parker. Adv. Eu.

Robinia, L. . . . . LOCUST.

R. Pseudacacia, Vent. Common Locust. Escaped from cultivation in many places. Adv. Southern and Western States.

R. viscosa, Vent. Clammy Locust. Sparingly escaped from cultivation. Princeton, O. R. Willis; Tom's River, N. L. Britton; Franklin, Essex Co., H. H. Rusby. Adv. Southern States.

Tephrosia, Pers. . . . . Hoary-pea.

T. Virginiana, Pers. Goat's Rue. Catgut. Common on the Yellow Drift, but rare north of it. Western bank of Greenwood Lake opposite Cooper's, W. H. Rudkin.

Desmodium, DC. . . . . Tick Trefoil.

D. nudiflorum, DC. Naked-flowered T. Dry woods. Quite common throughout.

D. acuminatum, DC. Naked-flowered T. Sparingly in the middle and northern parts of the State. Palisades, T. F. Allen; Snake Hill, N. L. Britton; Essex Co., H. H. Rusby.

D. rotundifolium, DC. Round-leaved T. Sandy or rocky woods. Most common in the northern counties.

D. canescens, DC. Hairy Tick Trefoil. Swampy ground. Not common. Banks of the Delaware near Gloucester, C. F. Parker; along First Mt., Essex Co., H. H. Rusby.

D. cuspidatum, Torr. & Gray. Large-bracted T. Not uncommon in the northern and middle counties. Chatham, W. H. Leggett.

D. lævigatum, DC. Smooth Tick Trefoil. Sparingly in the southern and middle counties. Pine barrens of Ocean Co., P. D. Knieskern; Bergen Point and Chatham, W. H. Leggett.

D. viridiflorum, Beck. Green-flowered T. Sparingly in the southern and central parts of the State.

D. Dillenii, Darlingt. Dillenius' Tick Trefoil. Open woods. Quite common throughout.

D. paniculatum, DC. Panicled Tick Trefoil. Copses and borders of woods. Common throughout.

D. strictum, DC. Erect Tick Trefoil. Confined to the area of the Yellow Drift, and quite common in the pine barrens. Woodbury, C. E. Smith; Malaga, Gloucester Co., C. F. Parker; rare in Ocean Co., P. D. Knieskern.

D. Canadense, DC. Canadian Tick Trefoil. Quite common in woods throughout the State.

D. ciliare, DC. Ciliate Tick Trefoil. Quite common on the Yellow Drift, but rare north of it.

D. Marylandicum, Boott. Maryland Tick Trefoil. Copses and open woods. Quite common throughout the State.

Lespedeza, Michx. . . . . Bush Clover.

L. repens, Bart. (Includes L. procumbens, Michx.) Creeping Bush Clover. Sandy woods and fields. Common in the southern and middle counties; sparingly in the northern parts of the State.

L. violacea, Pers. Violet Bush Clover. Dry fields and copses. Common throughout.

L. reticulata, Pers.; *Var.* angustifolia, Maxim. (**L.** violacea, Pers., *Var.* angustifolia, T. and G.) Sandy fields. Common in the southern and middle counties.

L. Stuvei, Nutt. Stuve's Bush Clover. "Along fences on hill-sides, Monmouth Co., common," O. R. Willis.

L. hirta, L. Hairy Bush Clover. Grows throughout the State, but is most abundant in the pine barrens.

L. capitata, Michx. Capitate Bush Clover. Dry sandy soil. Common throughout.

L. angustifolia, Ell. (**L.** capitata, Michx., *Var.* angustifolia, Pursh.) Narrow-leaved Bush Clover. Quite common on the Yellow Drift, but not elsewhere.

Stylosanthes, Swartz.   .   .   .   .   .   PENCIL-FLOWER.

S. elatior, Swartz. Pencil-flower. Middle and southern counties. Elizabethtown, Torrey Catalogue; South Amboy and Red Bank, W. H. Leggett; Monmouth and Ocean Cos., P. D. Knieskern; Cliffwood, Monmouth Co., W. H. Rudkin; New Brunswick, N. L. Britton; Milburn, Essex Co., W. M. Wolfe.

Coronilla, L.   .   .   .   .   CORONILLA.

C. varia, L. Common Coronilla. Plainfield, F. Tweedy; Guttenberg, G. M. Wilber. Adv. Eu.

Vicia, Tourn.   .   .   .   .   VETCH. TARE.

V. sativa, L. Common Vetch or Tare. Cultivated fields. Common throughout. Adv. Eu.

Var. angustifolia, Seringe. With the typical form in ballast at Camden, C. F. Parker. Adv. Eu.

V. tetrasperma, L. Four-seeded Vetch. Hoboken, Torrey Catalogue; Keyport, S. Lockwood; ballast at Camden, C. F. Parker; Communipaw, Addison Brown. Not common. Adv. Eu.

V. hirsuta, Koch. Hairy Vetch. About dwellings in eastern Monmouth Co., O. R. Willis; in ballast at Camden, C. F. Parker; and Communipaw, Addison Brown. Not common. Adv. Eu.

V. Cracca, L. Warren Co., F. Knighton; near Paterson, H. H. Rusby; in ballast at Camden, C. F. Parker; and Communipaw, A. Brown. Eu.

V. Caroliniana, Walt. Carolina Vetch. Hunterdon Co., T. C. Porter; Holland Station, Hunterdon Co., A. P. Garber. Rare.

V. Americana, Muhl. American Vetch. Probably grows in the northern counties, but no definite stations are reported.

Lathyrus, L. . . . VETCHLING. EVERLASTING-PEA.

L. maritimus, Bigelow. Beach Pea. Sea Bright, M. Ruger; shores
of New York Harbor, Torrey Catalogue. Rare. Eu.

L. palustris, L. Marsh Vetchling. New Durham, C. F. Austin;
banks of the Delaware near Camden, C. F. Austin. Rare. Eu.

Var. myrtifolius, Gray. Hackensack Marshes, W. H. Leggett; near
Phillipsburg, A. P. Garber; Kingsland Station, D. L. & W. R. R., H.
H. Rusby.

Apios, Boerh. . . . . GROUND-NUT. WILD BEAN.

A. tuberosa, Mœnch. Wild Bean. Low grounds. Common
throughout.

Phaseolus, L. . . . . KIDNEY BEAN.

P. perennis, Walt. Wild Bean. Warren Co., F. Knighton; Essex
Co., H. H. Rusby. Rare.

P. diversifolius, Pers. Wild Bean. Frequent along the sea-coast
and on sands of the Yellow Drift.

P. helvolvus, L. Wild Bean. Sandy fields. Quite common in the
southern and middle counties.

Clitoria, L.   . . . . BUTTERFLY-PEA.

C. Mariana, L. Butterfly Pea. Little Snake Hill, W. H. Leggett,
1871; Tom's River, P. D. Knieskern. Very rare.

Amphicarpæa, Ell. . . . . . . HOG PEA-NUT.

A. monoica, Nutt. Hog Pea-nut. Woods. Common throughout
the State, except in the pine barrens.

Galactia, P. Browne. . . . . '. MILK-PEA.

G. glabella, Michx. Smooth Milk-Pea. Rather frequent in the
pine barrens, and confined to the Yellow Drift.

Baptisia, Vent. . . . . FALSE INDIGO.

B. tinctoria, R. Br. Wild Indigo. Dry sandy soil. Quite common
throughout, but most abundant in the southern counties.

Cercis, L. . . . . . . RED-BUD. JUDAS-TREE.

C. Canadensis, L. Red-bud. Woods, New Jersey, Torrey Catalogue; in damp woods on bank of the Delaware River between Camden and Gloucester, C. F. Parker. Very rare.

Cassia, L. . . . . SENNA.

C. Marylandica, L. Wild Senna. Sparingly throughout the State.

C. Chamæcrista, L. Partridge Pea. Sandy fields and roadsides. Common in the southern and middle counties.

C. nictitans, L. Wild Sensitive-plant. Sandy fields and roadsides. Common in the southern and middle counties.

Gleditschia, L. . . . . HONEY LOCUST.

G. triacanthos, L. Honey Locust. Sparingly escaped from cultivation. Adv. Southwestern States.

ROSACEÆ.

Prunus, Tourn. . . . . PLUM. CHERRY.

P. Americana, Marsh. Wild Yellow or Red Plum. River banks and woods. Sparingly throughout the State.

P. maritima, Wang. Beach Plum. Sandy sea beaches, and occasionally on sandy soil a few miles inland.

P. spinosa, L. Sloe, Black-thorn. "Warren Co.," F. Knighton in Willis Catalogue. Adv. Eu.

P. pumila, L. Dwarf Cherry. Islands in the Delaware, above Phillipsburg, T. C. Porter ; Warren Co., F. Knighton ; Sussex Co., A. P. Garber. Rare and confined to rocky places in the northern counties.

P. Pennsylvanica, L. Wild Red Cherry. Sparingly in the northern and middle counties. Weehawken Heights, I. H. Hall ; Warren Co., F. Knighton ; Bergen Co., C. F. Austin ; Franklin, Essex Co., H. H. Rusby.

P. Virginiana, L. Choke-cherry. Sparingly in the northern and middle counties, growing along river banks. Sussex and Warren Cos., A. P. Garber ; near Closter, C. F. Austin.

P. serotina, Ehr. Wild Black Cherry. Open woods. Common throughout.

Spiræa, L. . . . . MEADOW-SWEET.

S. corymbosa, Raf. Corymbed Meadow-sweet. Near Chester, Morris Co., C. F. Austin.

S. salicifolia, L. Common Meadow-sweet, Low swampy ground. Rather common throughout the State. Eu.

S. tomentosa, L. Hardhack, Steeple-bush. Low swampy ground. Sparingly throughout.

Neillia, Don. (Spiræa, L.) . . . . . . NEILLIA.

N. opulifolia, Benth and Hook. Nine-bark. Rocky hills, New Jersey, Torrey Catalogue. Banks of Cooper's Creek, Camden, C. F. Parker. Rare.

Gillenia, Mœnch. . . . . INDIAN PHYSIC.

G. trifoliata, Mœnch. Bowman's Root. Rich woodlands. Rare. Bergen Co., C. F. Austin; Greenwood Lake, Jos. Schrenck ; Warren Co., C. F. Parker.

Poterium, L. . . . . BURNET.

P. Canadense, Benth and Hook. Canadian Burnet. Common in the northern and sparingly in the middle counties. Freehold, P. D. Knieskern; Plainfield, Frank Tweedy; Hightstown, O. R. Willis ; Snake Hill, W. M. Wolfe; New Durham, W. H. Leggett; Camden, C. F. Parker; Passaic Co., H. H. Rusby.

Agrimonia, Tourn. . . . . AGRIMONY.

A. Eupatoria, L. Common Agrimony. Woodlands. Rather common throughout the State, except in the pine barrens. Eu.

A. parviflora, L. Small-flowered A. Near Camden, C. F. Parker ; Plainfield, F. Tweedy; Closter, Bergen Co., C. F. Austin ; Essex Co., H. H. Rusby. Rare.

Geum, L. . . . . AVENS.

G. album, Gmelin. White Avens. Common in the northern and middle counties.

G. Virginianum, L. Virginian Avens. Sparingly in low grounds throughout the State.

G. strictum, Ait. Yellow Avens. Rare, and mostly confined to the northern counties. Sussex Co., A. P. Garber; Franklin, Essex Co., H. H. Rusby ; Parsippany, C. F. Austin ; " damp shady places, Ocean and Monmouth Cos., P. D. Knieskern (?);" Troy, Morris Co., C. F. Austin ; Long Hill, W. H. Leggett. Eu.

G. rivale, L. Water or Purple Avens. Sparingly in the northern counties. Near Closter, Bergen Co., and in Sussex Co., C. F. Austin ; Morris Co., A. P. Garber. Eu.

Waldsteinia, Willd.   . . . BARREN STRAWBERRY.

W fragarioides, Tratt. Barren Strawberry. Sparingly in the northern counties. Warren and Sussex Cos., O. R. Willis; Andover, Sussex Co., C. F. Austin.

Potentilla, L. . . . . CINQUE-FOIL. FIVE-FINGER.

P. Norvegica, L. Norwegian Five-finger. Common in the northern and middle counties. Rare on the Yellow Drift. Eu.

P. Canadensis, L. Common Five-finger. Dry soil. Common throughout.

Var. simplex, Torr. & Gray. Wet places. Common.

P. argentea, L. Silvery Cinque-foil. Barren fields. Sparingly in the northern and middle counties. Eu.

P. arguta, Pursh. Rocky places. Rare, and confined to the northern counties. On the Delaware below Phillipsburg, T. C. Porter; Cooper's Furnace, A. P. Garber.

P. anserina, L. Silver-weed. Sparingly in the northern and middle counties. Shores of Newark Bay, W. H. Leggett; ballast at Camden, C. F. Parker. Eu.

P. fruticosa, L. Shrubby Cinque-foil. Wet grounds. Frequent in the northern counties. Tenafly, Addison Brown; Sussex Co., A. P. Garber; Morris Co., C. F. Austin; Shippenport, Morris Co., H. H. Rusby; in meadows, Weehawken, Torrey Catalogue; Great Meadows, Warren Co., Prof. George H. Cook. Eu.

P. tridentata, Ait. Three-toothed Cinque-foil. Top of High Point, Sussex Co., C. F. Austin.

P. palustris, Scop. Marsh Five-finger. Budd's Lake, Morris Co., C. F. Parker, T. C. Porter. Very rare. Eu.

Fragaria, Tourn.   . . . . . . STRAWBERRY.

F. Virginiana, Ehr. Virginian Strawberry. Fields and open woods. Common throughout.

F. vesca, L. European Strawberry. Rather common in the northern counties. Rare elsewhere. Eu.

F. Indica, Andr. Indian Strawberry. Guttenberg, Hudson Co., M. Ruger; Ocean Co., C. F. Austin. Rare. Adv. India.

Rubus, Tourn. . . . . . RASPBERRY. BLACKBERRY.

R. odoratus, L. Purple-flowering Raspberry. Rocky places in the northern and middle counties. Frequent.

R. triflorus, Richardson. Dwarf Raspberry. Damp places; northern and middle counties. Rare. Bergen Co., C. F. Austin; New Durham, N. L. Britton; Monmouth Co., Dr. Torrey; Budd's Lake, Morris Co., T. C. Porter.

R. strigosus, Michx. Wild Red Raspberry. Sparingly in the northern parts of the State. Bergen Co., C. F. Austin; Warren Co., C. F. Parker; Franklin. Essex Co., H. H. Rusby.

R. occidentalis, L. Black Raspberry. Sparingly in the middle, and common in the northern counties.

R. villosus, Ait. High Blackberry. Fields and thickets. Common throughout.

R. Canadensis, L. Low Blackberry. Dewberry. Rocks and sandy fields. Common throughout.

R. hispidus, L. Running Swamp Blackberry. Swampy places. Rather common throughout.

R. cuneifolius, Pursh. Sand Blackberry. Common on the Yellow Drift and confined to that formation.

Rosa, Tourn. . . . . Rose.

R. Carolina, L. Swamp Rose. Low grounds. Common throughout.

R. lucida, Ehrhart. Dwarf Wild Rose. Dry fields and roadsides. Rather common.

R. blanda, Ait. Early Wild Rose. "Damp meadows, Freehold. Not common." O. R. Willis in Catalogus Plantarum. The only station mentioned in the State.

R rubiginosa, L. Sweet Brier. Roadsides and thickets. Sparingly throughout. Nat. Eu.

R. micrantha, Smith. Small-flowered Sweet Brier. Rare. Hoboken, C. F. Austin. Nat. Eu.

Cratægus, L. .   .   HAWTHORN. WHITE-THORN

C. Oxyacantha, L. English Hawthorn. Sparingly escaped from cultivation. Warren Co., F. Knighton; Hudson Co., C. F. Austin. Adv. Eu.

C. coccinea, L. Scarlet-fruited Thorn. Thickets and rocky banks. Frequent throughout the State.

C. tomentosa, L. Black Thorn, Pear Thorn. Sparingly in thickets in the northern counties. Verona and Caldwell, Essex Co., (the Var. pyrifolia, Gray), H. H. Rusby; also in Monmouth Co., R. W. Brown.

C. Crus-galli, L. Cockspur Thorn. Sparingly in thickets throughout the State.

C. parvifolia, Ait. Dwarf Thorn. Common on the area of the Yellow Drift and mostly confined to it; but grows also at Plainfield, Frank Tweedy; on the Palisades, C. F. Austin; and at Milford, Hunterdon Co., N. L. Britton.

Pirus, L. (Pyrus, L.) . . . . . , . Pear. Apple.

P. coronaria, L. American Crab Apple. Sparingly in the northwestern parts of the State. Warren Co., C. F. Parker; Morris Co., C. F. Austin.

P. arbutifolia, L. Chokoberry Damp Thickets. Common throughout.

Var. melanocarpa, Hook. Chokeberry. Damp thickets. Rather common throughout.

P. Americana, DC. American Mountain-ash. Very rare and confined to the northern counties. Budd's Lake, T. C. Porter.

Amelanchier, Medic. . . . . JUNE-BERRY.

A. Canadensis, Torr. and Gray. Shad Bush. Along streams and in low grounds. Quite common in the northern and middle counties; less so in the southern parts of the State.

Var. (?) oblongifolia, Torr. and Gray. Shad Bush. Similar situations, but much less common.

SAXIFRAGACEÆ.

Ribes, L. . . . . CURRANT. GOOSEBERRY.

R. Cynosbati, L. Wild Gooseberry. Rare, and confined to the northern counties. Closter, C. F. Austin; Preakness Mt., W. L. Fischer.

R. oxyacanthoides, L. Wild Gooseberry. Rare. Closter, C. F. Austin; Plainfield, Frank Tweedy.

R. rotundifolium, Michx. Wild Gooseberry. Rocky places in the northern counties. Warren Co., C. F. Parker; Fort Lee, Bergen Co., W. H. Leggett.

R. prostratum, L'Her. Fetid Currant. Very sparingly in the northern parts of the State. Closter, C. F. Austin.

R. floridum, L'Her. Wild Black Currant. Sparingly throughout. Princeton, Dr. John Torrey; Warren Co., F. Knighton; Morris Co., C. F. Austin; Snake Hill, M. Ruger.

R. rubrum, L. Red Currant. Sparingly escaped from cultivation into woods and thickets. New Durham, T. F. Allen; Plainfield, F. Tweedy; Camden Co., C. F. Parker. Eu.

Itea, L. . . . . ITEA.

I. Virginica, L. Virginian Itea. Swamps in the southern and southeastern counties, and mostly confined to the pine barrens. Manchester, Ocean Co., P. D. Knieskern; Tom's River, Jos. Schrenck.

Hydrangea, Gronov. . . . . HYDRANGEA.

H. arborescens, L. Wild Hydrangea. Rocky places in the northern parts of the State. Hunterdon and Warren Cos., C. F. Parker; Delaware Water Gap, H. H. Rusby; near Phillipsburg, T. C. Porter.

Parnassia, L.   . .   . . GRASS OF PARNASSUS.

P. Caroliniana, Michx. Grass of Parnassus. Sparingly in the northern and middle counties. Sussex Co., A. P. Garber; Closter, Bergen Co., C. F. Austin; marl banks, New Egypt, Ocean Co., O. R. Willis; Great Meadows, Warren Co., Prof. Geo. H. Cook.

4

Saxifraga, L. . . . . SAXIFRAGE.

S. Virginiensis, Michx.   Early Saxifrage.   Dry or rocky banks.
Very common in the northern and middle counties, but rare in the
pine barrens.

S. Pennsylvanica, L.   Swamp Saxifrage.   Rather common in bogs
in the middle and northern counties; rare on the Yellow Drift.
Middletown, Monmouth Co., R. W. Brown.

Heuchera, L. . . . . ALUM–ROOT.

H. Americana, L.   Common Alum-root.   Common on shady banks,
except in the pine barrens, where it is rarely found.

Mitella, Tourn. . . . MITRE-WORT. BISHOP'S-CAP.

M. diphylla, L.   Two-leaved Mitre-wort.   Sparingly in the northern
and middle counties.   Palisades and Closter, C. F. Austin; Warren
Co., C. F. Parker; Preakness Mt., W. L. Fischer; Plainfield, F.
Tweedy; Morristown and Hemlock Falls, Essex Co., W. H. Leggett;
Eagle Rock, Essex Co., W. M. Wolfe; Parsippany, Miss E. G. Knight.

Tiarella, L. . . . . FALSE MITRE-WORT.

T. cordifolia, L.   False Mitre-wort.   Among rocks at Passaic Falls,
Torrey Catalogue; Limestone rocks, Sussex Co., C. F. Austin.   Very
rare.

Chrysoplenium, Tourn. . . . GOLDEN SAXIFRAGE.

C. Americanum, Schwein.   Golden Saxifrage.   Wet places.   Common in the northern and middle parts of the State, but rare in the
pine barrens.

CRASSULACEÆ.

Sedum, Tourn. . . . . . . STONE-CROP. ORPINE.

S. ternatum, Michx.   Stone-crop.   Roadsides near Rockland, Bergen Co., C. F. Austin in Willis Catalogue.   Probably escaped from
cultivation.

S. Telephium, L.   Live-for-ever.   Roadsides.   Occasionally escaped
from cultivation.   Adv. Eu.

Penthorum, L. . . . . DITCH STONE-CROP.

P. sedoides, L.   Ditch Stone-crop.   Wet places.   Common throughout.

DROSERACEÆ.

Drosera, L. . . . . SUNDEW.

D. rotundifolia, L.   Round-leaved Sundew.   Peat-bogs.   Common
throughout the State.   Eu.

D. intermedia, Drev. and Hayne. *Var.* Americana, DC. (**D.** long-ifolia, L.) Peat-bogs. Common in the pine barrens, and sparingly throughout the rest of the State.

D. filiformis, Raf. Thread-leaved Sundew. Sandy swamps. Abundant in the pine barrens and confined to the Yellow Drift.

HAMAMELACEÆ.

Hamamelis, L. . . . . WITCH-HAZEL.

H. Virginica, L. Witch-hazel. Damp woods. Grows throughout the State, but is most abundant in the northern counties.

Liquidambar, L. . . . . . . SWEET-GUM TREE.

L. Styraciflua, L. Sweet-gum. Alligator Wood. Damp woods. Very common in the middle and southern counties, and frequent in the northern parts of the State.

HALORAGEÆ.

Myriophyllum, Vaill. . . . . WATER-MILFOIL.

M. scabratum, Michx. Water-milfoil. In ponds, but rare. Near Freehold, O. R. Willis; Cape May, C. F. Parker, W. M. Canby.

M. ambiguum, Nutt. *Var.* limosum, Torrey. Gloucester Co., C. F. Parker; Keyport, S. Lockwood.

Var. capillaceum, Torr. and Gray. Egg Harbor City, C. F. Parker.

Proserpinaca, L. . . . . . . MERMAID-WEED.

P. palustris, L. Common Mermaid-weed. Swamps. Rather common throughout.

P. pectinata, Lam. Pectinate Mermaid-weed. Sandy swamps. Rare. Manchester, Ocean Co., O. R. Willis; Atlantic City, W. M. Canby; Franklin, Essex Co., H. H. Rusby.

Callitriche, L. . . . . WATER-STARWORT.

C. Austini, Engelm. Austin's W. Damp soil. Rare. Closter, Bergen Co., C. F. Austin; Palisades, W. H. Leggett.

C. verna, L. Spring Water-starwort. Ponds and brooks. Rather common throughout. Eu.

C. heterophylla, Pursh. Various-leaved W. Ponds and brooks. Frequent.

Var. linearis, Pursh. Immersed and forming large floating masses in the Hackensack River, Bergen Co., C. F. Austin.

MELASTOMACEÆ.

Rhexia, L. . . . . MEADOW-BEAUTY.

R. Virginica, L. Meadow-beauty. Sandy swamps. Common in the middle and southern counties.

R. Mariana, L. Meadow Beauty. Sandy swamps. Rare, and confined to the southern counties.

LYTHRACEÆ.

Ammannia, Houston. . . . . AMMANNIA.

A. humilis, Michx. Low Ammannia. Damp places; rare. Closter, Bergen Co., C. F. Austin; Hackensack Meadows, Torrey Catalogue; Camden, C. F. Parker.

Lythrum, L. . . . . LOOSESTRIFE.

L. Hyssopifolia, L. Hyssop-leaved L. Marshes along the coast, Gray's Manual. Eu.

L. lineare, L. Linear-leaved L. Hackensack Meadows, T. F. Allen, W. H. Leggett; near Little Snake Hill, J. W. Congdon; borders of salt marshes in Monmouth, Ocean, and Middlesex Cos., O. R. Willis; Keyport, R. W. Brown.

L. Salicaria, L. Spiked Loosestrife. Banks of the Delaware River, Pavonia, and in ballast at Camden, C. F. Parker; Hudson Co., C. F. Austin; near Granton, N. R. R. of N. J., W. M. Canby; Plainfield, Frank Tweedy. Adv. Eu.

Nesæa, Commerson. . . . . SWAMP LOOSESTRIFE.

N. verticillata, H. B. K. Swamp Loosestrife. Swamps; quite common throughout.

Cuphea, Jacq. . . . . CUPHEA.

C. viscosissima, Jacq. Clammy Cuphea. Dry fields. Not common. Plainfield, F. Tweedy; Closter, C. F. Austin; near Camden, C. F. Parker; near Keyport, R. W. Brown; Franklin, Essex Co., H. H. Rusby.

ONAGRACEÆ.

Circæa, Tourn. . . . . ENCHANTER'S NIGHTSHADE.

C. Lutetiana, L. Enchanter's Nightshade. Common in the northern and middle counties. Eu.

Gaura, L. . . . .   GAURA.

G. biennis, L. Biennial Gaura. Banks. Camden, C. F. Parker. Rare.

Epilobium, L. . . . .   WILLOW-HERB.

E. spicatum, Lam. (**E.** angustifolium, L.) Great Willow-herb. Low grounds; rather common throughout. Eu.

E. palustre, L.; *Var.* lineare, Gray. Swamps in the northern counties. Rather rare. Closter, C. F. Austin; Budd's Lake, Morris Co., T. C. Porter; Sussex Co., C. F. Austin; near Passaic and Franklin, Essex Co., H. H. Rusby. Eu.

E. molle, Torrey. Downy Willow-herb. Morristown and Chatham, rare, W. H. Leggett.

E. coloratum, Muhl. Swamp Willow-herb. Swamps. Common throughout the State.

Œnothera, L. . . . .   EVENING PRIMROSE.

Œ. biennis, L. Common Evening Primrose. Dry fields. Common throughout.

Var. muricata, Lindl. Dry fields. Common throughout.

Œ. humifusa, Nutt. Drifting sand at Cape May, C. F. Parker.

Œ. sinuata, L. Abundant in the sands of the pine barrens, and confined to the Yellow Drift area.

Var. minima, Nutt. Frequent with the type.

Œ. pumila, L. (Includes **Œ.** chrysantha, Michx.) Sparingly in the northern and middle counties. Mt. north of Closter, C. F. Austin; Long Hill, W. H. Leggett; Plainfield, F. Tweedy; Franklin, Essex Co., H. H. Rusby; near Old Bridge, Middlesex Co., R. W. Brown.

Œ. fruticosa, L. Sundrops. Dry fields. Common throughout.

Var. linearis, Watson. (**Œ.** riparia, Nutt.) Ocean and Cape May counties, C. F. Parker; meadows near Plainfield, F. Tweedy; Camden, W. M. Canby; Quaker Bridge, Dr. Asa Gray.

Var. humifusa, T. F. Allen. Ocean Grove, I. Burk. (?)

Ludwigia, L. . . . .   FALSE LOOSESTRIFE.

L. alternifolia, L. Seed-box. Swamps. Quite common throughout.

L. hirtella, Raf. Sparingly in wet places in the pine barren region.

L. sphærocarpa, Ell. Rare. Closter, Bergen Co., C. F. Austin; Atsion, Burlington Co., C. F. Parker.

L. linearis, Walt. Bogs in the pine barrens. Not common.

L. palustris, Ell. Ponds and ditches. Common throughout. Eu.

CUCURBITACEÆ.

Echinocystis, Torr. & Gray. .   WILD BALSAM–APPLE,

E. lobata, Torr. & Gray. Wild Balsam-apple. Rare. Near Burlington, Isaac Burk.

Sicyos, L. . . . . . ONE-SEEDED STAR CUCUMBER.

S. angulatus, L. One-seeded Cucumber. Sparingly in damp places throughout the State.

CACTACEÆ.

Opuntia, Tourn. . . . . PRICKLY PEAR.

O. vulgaris, Haworth. Prickly Pear. Rare. Haddonfield, I. C. Martindale; South Jersey, W. M. Canby.

O. Rafinesquii, Engl. Prickly Pear. Sandy fields and on rocks. Frequent throughout the State.

FICOIDEÆ.

Mollugo, L. . . . . INDIAN–CHICKWEED.

M. verticillata, L. Carpet-weed. Waste and cultivated grounds. Common. Adv. Southern States.

Sesuvium, L. . . . . SEA PURSLANE.

S. pentandrum, Ell. Sea Purslane. Frequent on the coast from Sandy Hook to Cape May.

UMBELLIFERÆ.

Hydrocotyle, L. . . . . . WATER PENNYWORT.

H. Americana, L. Banks of Shark River, P. D. Knieskern; Keyport, R. W. Brown; near Freehold, S. Lockwood; and common in the northern parts of the State.

H. umbellata, L. Frequent in the southern and middle counties. Atlantic City, Cape May and Camden, C. F. Parker; Point Pleasant, Ocean Co., P. D. Knieskern; Red Bank, along the muddy shore of the Navesink River, W. H. Leggett.

Var. (?) ambigua, Gray. Cape May, C. F. Parker.

H. interrupta, Muhl. Sparingly near the coast. Red Bank, W. H. Leggett; Cape May, W. M. Canby.

Eryngium, L. .   . . . ERYNGO.

E. yuccæfolium, Michx. "Pine barrens." Gray's Manual and C. F. Austin, but no definite stations are reported.

E. Virginianum, Lam. In swamps. Frequent along the coast, and mostly confined to the Yellow Drift area. Spring Lake, Monmouth Co., Addison Brown; borders of salt meadows at Hoboken, Torrey Catalogue.

Sanicula, Tourn. . . . . . . BLACK SNAKE-ROOT.

S. Canadensis, L. Sanicle. Woods and copses. Common except in the pine barrens.

S. Marylandica, L. Sanicle. Common in similar situations, and having the same range as the last species.

Daucus. Tourn. . . . . CARROT.

D. carota, L. Wild Carrot. Meadows. Too common. Nat. Eu.

Heracleum, L. . . . . COW-PARSNIP.

H. lanatum, Michx. Wooly Cow-parsnip. Mercer Co., Dr. John Torrey; Camden, C. F. Parker; borders of salt meadows, Hoboken, Torrey Catalogue, and sparingly in the northern counties.

Pastinaca, Tourn. . . . . PARSNIP.

P. sativa, L. Common Parsnip. Fields and roadsides. Common in all parts of the State. Adv. Eu.

Archemora, DC. . . . . COWBANE.

A. rigida, DC. Cowbane. Northern R. R. of N. J., T. F. Allen, C. F. Austin; Fairfield, Torrey Catalogue; New Brooklyn, Middlesex Co., F. Tweedy; and frequent in sandy swamps in the Yellow Drift area.

Var. ambigua, Torr. and Gray. Quaker Bridge, Burlington Co., C. F. Parker.

Archangelica, Hoffm. . . . . ARCHANGELICA.

A. hirsuta, Torr. and Gray. Hairy Angelica. Sandy woods. Common in the middle and southern counties. Essex Co., H. H. Rusby.

A. atropurpurea. Hoffm. Great Angelica. Sparingly in swamps in the northern and middle counties. Closter, Bergen Co., C. F. Austin; N. R. R. of N. J., W. H. Leggett; Plainfield, Frank Tweedy.

Selinum, L. (Conioselinum, Fisch.) . MILK-PARSLEY.

S. Canadense, Michx. Canadian Milk-parsley. Closter, Bergen Co., C. F. Austin. Rare.

Æthusa, L. . . . . Fool's Parsley.
Æ. Cynapium, L. Fool's Parsley. Near Pleasant Valley on road
to Fort Lee, Bergen Co., W. H. Leggett; waste places, Plainfield, F.
Tweedy; Haddonfield, and in ballast at Camden, C. F. Parker. Adv.
Eu.

Thaspium, Nutt. . . . . Meadow-parsnip.
T. barbinode, Nutt. Hunterdon Co., A. P. Garber; Princeton, Mer-
cer Co., O. R. Willis; "shady banks, Prospertown, Ocean Co., rare,"
P. D. Knieskern
T. aureum, Nutt. Low grounds. Sparingly throughout the State.
Pascack and Weehawken, C. F. Austin; Ocean Co., P. D. Knieskern;
banks of the Delaware, Gloucester Co., C. F. Parker.
T. trifoliatum, Gray. Rocky woodlands in the middle and north-
ern counties. Long Hill, W. H. Leggett; Weehawken, C. F. Austin;
common in Essex Co., H. H. Rusby.
Var. atropurpureum, Torr. & Gray. Camden Co., C. F. Parker;
Stony Brook, Plainfield, Frank Tweedy.

Pimpinella, L. (Zizia, L.) . . . . . Wild Zizia.
P. integerrima, Benth. & Hook. Rocky hillsides, middle and
northern counties. Near English Neighborhood, C. F. Austin; Palisades, T. F. Allen; rare in Monmouth and Ocean Cos., P. D. Knieskern; Long Hill, W. H. Leggett; Plainfield, Frank Tweedy; Warren
Co., A. P. Garber.

Bupleurum, Tourn. . . . . Thorough-wax.
B. rotundifolium, L. Thorough-wax. Rare. Mercer Co., Dr. John
Torrey; Woodbury, W. M. Canby. Adv. Eu.

Discopleura, DC. . . . . . . Mock Bishop-weed.
D. capillacea, DC. Common along the coast; usually, but not
always, growing in brackish swamps.

Cicuta, L. . . . . , . . Water-hemlock.
C. maculata, L. Spotted Cowbane. Swamps. Common except in
the pine barrens.
C. bulbifera, L Bulb-bearing Water-hemlock. Swamps in the
northern and middle counties. Salt marshes, Hoboken, Torrey Cata-
logue; Closter, C. F. Austin; Plainfield, Frank Tweedy; Budd's Lake,
Morris Co., T. C. Porter; Sussex Co., A. P. Garber; Franklin, Essex
Co., H. H. Rusby; Camden Co., C. F. Parker.

Sium, L. . . . . WATER PARSNIP.

S. cicutæfolium, Gmel. (S. lineare, Michx.) Water Parsnip. Quite common throughout, except in the pine barren region.

Cryptotænia, DC. . . . . HONEWORT.

C. Canadensis, DC. Canadian Honewort. Common in the northern and middle counties; rare on the Yellow Drift.

Chærophyllum, L. . . . . CHERVIL.

C. procumbens, Lam. Low Chervil. Hoboken Hills, Torrey Catalogue; South Jersey, rare, C F. Austin; banks of the Delaware, near Camden, C. F. Parker. Rare.

Osmorrhiza, Raf. . . . . SWEET CICELY,

O. longistylis, DC. Smoother Sweet Cicely. Sparingly in the middle and northern counties. Closter, Bergen Co., C. F. Austin; Long Hill and Chatham, W. H. Leggett; three miles above Newark, on the Passaic River, I. H. Hall; near Keyport, R. W. Brown.

O. brevistylis, DC. Hairy Sweet Cicely. Rather common throughout, but most abundant in the northern counties.

Conium, L. . . . . , ... POISON HEMLOCK.

C. maculatum, L. Poison Hemlock. Sussex Co., C. F. Austin; Mercer Co., Dr. John Torrey; Bool's Island, in Delaware River, I. S. Moyer; Phillipsburg, T. C. Porter; Essex Co., H. H. Rusby. Adv. Eu.

ARALIACEÆ.

Aralia, L. . . . . GINSENG. WILD-SARSAPARILLA.

A. spinosa, L. Angelica-tree. Hercules' Club. Sparingly escaped from cultivation. Plainfield, F. Tweedy.

A. racemosa, L. Spikenard. Rich woodlands; frequent in the middle and northern counties. Marble Hill above Phillipsburg, Warren Co., T. C. Porter; Closter, Bergen Co., C. F. Austin; Long Hill, W. H. Leggett; near Holmdel, Monmouth Co., R. W. Brown; Camden Co., C. F. Parker; First Mt., Essex Co., H. H. Rusby; Plainfield, F. Tweedy.

A. hispida, Michx. Bristly Sarsaparilla. Rocky places in the northern counties, rare; near Lodi Junction, N. J. and N. Y. R. R., W. H. Rudkin; also, in Secaucus Swamp, W. H. Leggett; and in sandy pine barrens of Ocean Co., P. D. Knieskern! !

A. nudicaulis, L. Wild Sarsaparilla. Frequent throughout the northern and middle counties. Rare on the Yellow Drift.

A. quinquefolia, Decsne. and Planch. Ginseng. Plainfield, Frank Tweedy. The only known locality in the State.

A. trifolia, Decsne. and Planch. Dwarf Ginseng. Sparingly throughout the northern and middle counties. Near Freehold, O. R. Willis; Closter, Bergen Co., and on rocks of Sussex Co., C. F. Austin; Camden Co., C. F. Parker; common in Essex Co., H. H. Rusby; Marble Hill, above Phillipsburg, T. C. Porter; abundant at Cranford, C. R. R. of N. J., N. L. Britton; Panrapo woods, W. H. Rudkin; Succasunna, Morris Co., T. C. Porter.

CORNACEÆ.

Cornus, Tourn. . . . . . . CORNEL. DOGWOOD.

C. Canadensis, L. Dwarf Cornel. Rare, and confined to the northern parts of the State. New Durham Swamp, Torrey Catalogue, 1819, and C. F. Austin, 1861.

C. florida, L. Flowering Dogwood. Open woods. Common throughout the State.

C. circinata, L'Her. Round-leaved Cornel. Rocky places; sparingly in the northern and middle counties. Closter, Bergen Co., C. F. Austin; Sussex Co., A. P. Garber; Warren Co., C. F. Parker; Plainfield, F. Tweedy.

C. sericea, L. Silky Cornel. Kinnikinnik. Frequent in the middle and northern counties. Rare on the Yellow Drift.

C. stolonifera, Michx. Red-osier Dogwood. Frequent, except in the southern parts of the State.

C. paniculata, L'Her. Panicled Cornel. Frequent, except in the southern parts of the State.

C. alternifolia, L. Alternate-leaved Cornel. Frequent, except in the southern parts of the State.

Nyssa, L. . . . TUPELO. PEPPERIDGE. SOUR GUM.

N. multiflora, Wang. Black or Sour Gum. Common throughout the State.

Division B.—Gamopetalæ.

CAPRIFOLIACEÆ.

Linnæa, Gronov. . . . . TWIN-FLOWER.

L. borealis, Gronov. Twin-flower. New Durham Swamp, 1861, C. F. Austin; near Paterson, Wm. Bower. Eu.

Lonicera, L. . . .   . WOODBINE. HONEYSUCKLE.

L. sempervirens, Ait. Trumpet Honeysuckle. Rather rare, and mostly confined to the middle and southern counties. Gloucester Co., C. F. Parker; Plainfield, Frank Tweedy ; New Durham Swamp, W. H. Leggett; rare in Essex Co., H. H. Rusby; near Princeton, O. R. Willis.

L. grata, Ait. American Woodbine. Rare, and confined to the northern counties. New Durham Swamp, Torrey Catalogue; Warren Co., F. Knighton.

L. parviflora, Lam. Small Honeysuckle. Sparingly in the middle and northern counties, growing on rocks. Closter, Bergen Co., C. F. Austin; Preakness, Passaic Co., W. L. Fischer; Palisades and Secaucus, W. H. Leggett, N. L. Britton; Plainfield, F. Tweedy; eastern Essex Co., H. H. Rusby; Marble Hill near Phillipsburg, T. C. Porter.

Diervilla, Tourn. . . . . BUSH HONEYSUCKLE.

D. trifida, Mœnch. Bush Honeysuckle. Sparingly on rocks in the northern counties. Warren Co., C. F. Parker: Long Hill, W. H. Leggett; Preakness, W. L. Fischer : Verona, Essex Co., H. H. Rusby; Stanhope, Morris Co., T. C. Porter.

Triosteum, L. . . . FEVER-WORT. HORSE-GENTIAN.

T. perfoliatum, L. Horse-gentian. Frequent in the northern and middle counties; rare on the Yellow Drift.

Sambucus, Tourn. . . . . ELDER.

S. Canadensis, L. Common Elder. Rich soil. Common, except in the pine barren regions.

S. pubens, Michx. Red-berried Elder. Sparingly in rocky places in the northern and middle counties. Palisades, C. F. Austin; at Fort Lee, N. L. Britton; Marble Hill above Phillipsburg, Warren Co., T. C. Porter; Bool's Island, Delaware River, I. S. Moyer ; on First Mt., Essex Co., H. H. Rusby.

Viburnum. L.   . . ARROW-WOOD. LAURESTINUS.

V. Lentago, L. Sweet Viburnum. Sheep-berry. Rather common in the northern and middle counties.

V. prunifolium, L. Black Haw. Nanny-berry. Common, except in the pine barrens.

V. nudum, L. Withe-rod. Rather common in swamps.

Var. Claytoni, Gray. Clayton's Viburnum. Sandy swamps. Common in the pine barrens; also in Secaucus Swamp, W. H. Leggett.

V. dentatum, L. Arrow-wood. Swamps. Common throughout the State.

V. pubescens, Pursh. Downy Arrow-wood. Sparingly in rocky places in the middle and northern counties. Hills near Princeton, O. R. Willis; Closter, Bergen Co., C. F. Austin; Preakness, Passaic Co., W. L. Fischer.

V. acerifolium, L. Maple-leaved Arrow-wood. Common in woods in the northern and middle counties; rare on the Yellow Drift. Near Keyport, R. W. Brown.

V. Opulus, L. Cranberry-tree. Very rare. Sussex Co., A. P. Garber. Eu.

RUBIACEÆ.

Galium, L. . . . . Bedstraw. Cleavers.

G. aparine, L. Cleavers. Goose-grass. Moist thickets. Common in the middle and northern counties. Adv. Eu. (?)

G. asprellum, Michx. Rough Bedstraw. Frequent in the northern counties. Rare elsewhere.

G. trifidum, L. Small Bedstraw. Common throughout the State. Eu.

G. triflorum, Michx. Sweet-scented Bedstraw. Common throughout the middle and northern counties; rare on the Yellow Drift. Eu.

G. pilosum, Ait. Hairy Bedstraw. Common on the Yellow Drift, and sparingly in other parts of the State.

Var. puncticulosum, Gray. Egg Harbor City, I. C. Martindale.

G. hispidulum, Michx. Cape May near the Landing, A. Commons.

G. circæzans, Michx. Wild Liquorice. Common, except in the pine barrens.

G. lanceolatum, Torr. Wild Liquorice. Sparingly in the northern and occasional in the middle counties. Closter, Bergen Co., C. F. Austin; Long Hill, W. H. Leggett; Warren Co., C. F. Parker; Verona, Essex Co., H. H. Rusby; Freehold and Hightstown, O. R. Willis.

G. boreale, L. Northern Bedstraw. Marble Hill above Phillipsburg, T. C. Porter; Chatham Station, M. & E. R. R., W. H. Leggett; Princeton, Dr. John Torrey. Rare, and mostly confined to the northern counties. Eu.

G. verum, L. Yellow Bedstraw. In ballast at Communipaw, N. L. Britton. Adv. Eu.

Diodia, L. . . , ... Button-weed.

D. Virginica, L. Virginian Button-weed. Cape May, C. F. Parker.

D. teres, Walt. Terete-stemmed B. Sandy fields and roadsides. Very common on the Yellow Drift, and sparingly elsewhere.

Cephalanthus, L. . . . . Button-bush.

C. occidentalis, L. Button-bush. Swamps, etc. Common throughout.

Mitchella, L. . . . . PARTRIDGE-BERRY.

M. repens, L.   Partridge-berry. Woods and copses.   Common throughout.

Oldenlandia, Plum., L. . . . . . . OLDENLANDIA.

O. glomerata, Michx.   Sparingly throughout the State.   Closter, Bergen Co., and Manchester, Ocean Co., C. F. Austin ; Camden Co., E. Diffenbaugh ; Atlantic City, C. F. Parker.

Houstonia, L. . . . . HOUSTONIA.

H. purpurea, L.   Purplish Houstonia.   Rare.   New Jersey, Torrey Catalogue.

H. cærulea, L.   Bluets.   Near Shark River, P. D. Knieskern ; near Paterson, C. F. Austin ; Warren Co., C. F. Parker ; northwestern Essex Co., H. H. Rusby ; Trenton, S. Lockwood ; Little Falls, Passaic Co., W. M. Wolfe ; Camden, W. M. Canby.   Mostly confined to the northern parts of the State.

VALERIANACEÆ.

Fedia, Gærtn. . . . CORN SALAD.   LAMB-LETTUCE.

F. olitoria, Vahl.   Lamb-lettuce.   Canal banks at Trenton, E. A. Apgar.   Adv. Eu.

F. radiata, Michx.   Corn Salad.   Gloucester Co., C. F. Parker.

DIPSACEÆ.

Dipsacus, Tourn. . . . . TEASEL.

D. sylvestris, Mill.   Wild Teasel.   Sparingly along roadsides, etc. Ocean Co., P. D. Knieskern ; Warren Co., F. Knighton ; Camden Co., and in ballast at Camden, C. F. Parker.   Nat. Eu.

D. Fullonum, L.   Fuller's Teasel.   Along the Passaic River in Essex Co., H. H. Rusby.   Adv. Eu.

COMPOSITÆ.

Vernonia, Schreb. . . . . IRON-WEED.

V. Noveboracensis, Willd.   Iron-weed.   Wet places.   Common throughout.   Occasionally exhibits albinism in the flowers.

Sclerolepis, Cass. . . . . SCLEROLEPIS.

S. verticillata, Cass.   Whorled Sclerolepis.   Rare and confined to the pine barren regions.   South Jersey, W. M. Canby ; Ocean, Burlington and Cape May Cos., C. F. Parker.

Liatris, Schreb. . . . . Button Snake-root.

L. scariosa, Willd. Button Snake-root. Swamps, N. J., Eddy in Torrey Catalogue; near Keyport, R. W. Brown; Midland R. R., east of Newfoundland Station, W. H. Rudkin. Rare.

L. spicata, Willd. Salt meadows, near Squan, Monmouth Co., and Point Pleasant, Ocean Co., P. D. Knieskern; Brownsville, Middlesex Co., R. W. Brown; Griffith's, Camden Co., C. F. Parker; Midland R. R., east of Newfoundland Station, W. H. Rudkin; half a mile west west of Norwood, Bergen Co., C. F. Austin; Morris Co., W. H. Leggett; Hackensack Meadows, W. M. Wolfe. Not common.

L. graminifolia, Willd.; *Var.* dubia, Gray. Common in the pine barren regions, and confined to the Yellow Drift.

Kuhnia, L. . . . . Kuhnia.

K. eupatorioides, L. Limestone rocks, Sussex Co., C. F. Austin; Milford, Hunterdon Co., A. P. Garber; Camden, W. M. Canby. Rare.

Eupatorium, Tourn. . . . . . . Thoroughwort.

E. purpureum, L. Joe-Pye Weed. Common throughout.

E. fœniculaceum, Willd. In ballast at Camden, C. F. Parker. Adv. Southern States.

E. leucolepis, Torr. and Gray. Sparingly in the pine barrens.

E. hyssopifolium, L. Frequent on the Yellow Drift, and confined to that formation.

E. album, L. Sparingly in the pine barrens, and confined to the Yellow Drift. Near Keyport, R. W. Brown; South River, W. H. Leggett; Navesink Highlands, Addison Brown.

E. teucrifolium, Willd. Common in the southern and frequent in the middle counties. Near Snake Hill, W. M. Wolfe; Plainfield, F. Tweedy; Keyport, R. W. Brown; Franklin, Essex Co., H. H. Rusby. Mostly confined to the Yellow Drift.

E. rotundifolium, L. Common on the Yellow Drift and rare north of it. Franklin, Essex Co., H. H. Rusby.

E. pubescens, Muhl. Sparingly, with the same range as the last species. Spring Lake, Monmouth Co., Addison Brown; Manasquan, O. R. Willis.

E. sessilifolium, L. Upland Boneset. Sparingly in rocky places, northern and middle counties. English neighborhood, Palisades, C. F. Austin; along First Mt., Essex Co., H. H. Rusby; Weehawken, W. H. Rudkin; Snake Hill, N. L. Britton.

E. resinosum, Torr. Frequent in the pine barrens and rarely found out of them. Confined to the Yellow Drift.

E. perfoliatum, L. Boneset. Low grounds. Common throughout.

E. ageratoides, L. White Snake-root. Rich woods. Middle and
northern counties. Frequent.

E. aromaticum, L. White Snake-root. Sparingly in the middle
and southern counties. Near Squan Village, Monmouth Co., P. D.
Knieskern ; Freehold, O. R. Willis ; Camden, C. F. Parker ; Plainfield,
Frank Tweedy.

  Mikania, Willd. . . . . . Climbing Hemp Weed.

M. scandens, L. Climbing Hemp Weed. Wet places. Rather
common throughout.

  Conoclinum, DC. . . . . Mist–flower.

C. cœlestinum, DC. Mist-flower. Rare. Cape May, C. F. Parker.

  Tussilago, Tourn. . . . . Coltsfoot.

T. Farfara, L. Coltsfoot. Wet places, Ocean and Monmouth Cos.,
rare, P. D. Knieskern. Nat. Eu.

  Sericocarpus, Nees. . . . White-topped Aster.

S. solidagineus, Nees. Closter, Bergen Co., C. F. Austin ; Ocean
and Monmouth Cos., P. D. Knieskern ; Camden Co., C. F. Parker.
Not common.

S. conyzoides, Nees. Dry open woods. Common throughout.

  Aster, L. . . . . Starwort. Aster.

A. corymbosus, Ait. Common in woods, northern and middle
counties ; rare on the Yellow Drift.

A. macrophyllus, L. Timber Creek, Camden Co., C. F. Parker ;
Colt's Neck, Monmouth Co., O. R. Willis ; Montclair, Essex Co., H.
H. Rusby ; Chatham, W. H. Leggett. And frequent in the northern
parts of the State.

A. Radula, Ait. Sparingly in the pine barrens and on the Yellow
Drift. Camden Co., C. F. Parker ; Mercer Co., O. R. Willis.

A. surculosus, Michx. ; and *Var.* gracilis, Gray. Rare, and confined to pine barren regions.

A. spectabilis, Ait. Common in the pine barrens, and confined to
the Yellow Drift.

A. concolor, L. Frequent in the pine barrens, and confined to the
Yellow Drift.

A. patens, Ait. Dry soil. Common throughout.

Var. phlogifolius, Gray. Weehawken and Long Hill, W. H. Leggett,

A. lævis, L. ; *Var.* lævigatus, Gray. Rather common in the northern and middle counties.

A. undulatus, L. Common in the northern and middle counties ;
rare on the Yellow Drift.

A. cordifolius, L. Woodlands. Common throughout.

A. sagittifolius, Willd. Sparingly in the northern and middle counties. Snake Hill, T. F. Allen; Summit, Union Co., W. H. Leggett; First Mt., Essex Co., H. H. Rusby; Princeton, Dr. John Torrey; Morris Co., C. F. Austin.

A. puniceus, L.; and *Var.* vimineus, Gr. Swamps. Common in the middle and northern counties.

A. Novæ-angliæ, L. Low grounds. Common in the middle and northern counties, and sparingly on the Yellow Drift. Tom's River, C. F. Parker; near Chesquake Creek, Middlesex Co., R. W. Brown.

A. ericoides, L. Rather common throughout.

A. multiflorus, Ait. Sparingly throughout. Communipaw, W. H. Leggett; Camden Co., C. F. Parker; Monmouth Beach Centre, A. Brown.

A. dumosus, L. Rather common in the middle and southern counties.

A. Tradescanti, L. Common throughout.

A. miser, L. Ait. Dry fields. Quite common throughout.

A. simplex, Willd. Low grounds. Frequent throughout.

A. tenuifolius, L. Closter, Bergen Co., C. F. Austin.

A. longifolius, Lam. Low grounds. Rather common and very variable.

A. prenanthoides, Muhl. Sussex Co., C. F. Austin.

A. acuminatus, Michx. Sparingly in the northern parts of the State. Near Closter, C. F. Austin; Essex Co., H. H. Rusby.

A. nemoralis, Ait. Common in bogs in the pine barrens; also, New Durham Swamp, Torrey Catalogue.

A. flexuosus. Nutt. Common in salt marshes.

A. linifolius, L. Common in salt marshes; also in ballast at Camden, C. F. Parker.

A. linariifolius, Hook. (Diplopappus, Cass.) Dry soil; quite common throughout. Not found in Essex Co., H. H. Rusby.

A. umbellatus, Torr. and Gray. (Diplopappus, Cass.) Rather common in swamps, middle and northern counties.

A. amygdalinus, Torr. and Gray. (Diplopappus, Cass.) New Jersey, Gray's Manual; low grounds, Monmouth and Ocean Cos., P. D. Knieskern. Rare.

A. cornifolius, Darl. (Diplopappus, Cass.) Frequent throughout the State. Chatham, W. H. Leggett; Franklin, Essex Co., H. H. Rusby; Monmouth and Ocean Cos., P. D. Knieskern.

Solidago, L . . . . GOLDEN-ROD.

S. squarrosa, Muhl. Sparingly in the northern parts of the State. Palisades, C. F. Austin; opposite Yonkers, N. Y., E. P. Bicknell.

S. bicolor, L. Dry open woods. Common throughout.

Var. concolor, Gray. Plainfield, I. H. Hall.

S. latifolia, L. Sparingly in the middle and northern counties. Camden, C. F. Parker; Plainfield, F. Tweedy; Closter, Bergen Co., C. F. Austin; New Durham Swamp, W. H. Leggett; not rare in Essex Co., H. H. Rusby.

S. cæsia, L. Common in the northern and middle counties.

S. virgata, Michx. Quite common in the pine barrens, and confined to the Yellow Drift.

S. puberula, Nutt. Frequent in the pine barrens, and mostly confined to the Yellow Drift. Sandy fields near Amboy, Nuttall in Torrey Catalogue; near Keyport, R. W. Brown; Chatham, W. H. Leggett.

S. stricta, Ait. Budd's Lake, Morris Co., T. C. Porter; Warren Co., F. Knighton; in a swampy bog, Succasunna, Morris Co., C. F. Austin. Rare, and confined to the northern counties.

S. speciosa, Nutt. Rare. Chatham, W. H. Leggett; Palisades, opposite Yonkers, N. Y., E. P. Bicknell; Montclair Heights near the base of the mountain, opposite R. R. Station, W. H. Rudkin.

Var. angustata, Gray. Pine barrens of Atlantic Co., C. F. Parker.

S. rigida, L. Rare. Palisades, C. F. Austin; Cooper's Furnace, Sussex Co., A. P. Garber; Little Snake Hill, W. H. Leggett; Warren Co., F. Knighton.

S. sempervirens, L. Common in salt marshes.

S. elliptica, Ait. Hackensack Meadows near New York, John Carey; Brown's Mills, Burlington Co., C. F. Parker. Rare.

S. neglecta, Torr. and Gray. Bergen and Morris Cos., C. F. Austin; Hackensack Swamps, and South River, W. H. Leggett; Camden Co., C. F. Parker. Not common.

S. patula, Muhl. Carlstadt, Chatham and New Durham, W. H. Leggett; Closter, Bergen Co., C. F. Austin; Freehold, O. R. Willis; Budd's Lake, Morris Co., T. C. Porter; Franklin, Essex Co., H. H. Rusby.

S. arguta, Ait.; and *Var.* juncea, Gray. Rather common in the northern counties and near New York.

S. Muhlenbergii, Torr. and Gray. Sparingly in the northern and middle counties. Sussex Co., C. F. Austin; near Cooper's Furnace, A. P. Garber; Chatham and New Providence, W. H. Leggett.

S. linoides, Solander. Sparingly in bogs in the pine barrens. Tom's River, C. F. Parker, P. D. Knieskern; Ferrago, Ocean Co., C. F. Austin.

S. altissima, L. Fields and copses. Common throughout.

S. ulmifolia, Muhl. Woods and copses. Rather common throughout.

S. pilosa, Walt. Sparingly in the pine barrens, and confined to the Yellow Drift.

S. odora, Ait. Common in the southern and middle counties; scarce northward.

S. nemoralis, Ait. Very common throughout.

S. Canadensis, L. Fields and copses. Common throughout.

Var. procera, Gray. Hackensack Swamps, T. F. Allen.

S. serotina, Ait. Swamps and low grounds. Frequent.

S. gigantea, Ait. Fields and copses. Common throughout.

S. lanceolata, L. Fields and pastures. Common throughout.

S. tenuifolia, Pursh. Common on the Yellow Drift, and sparingly north of that formation. New Durham Swamp, C. F. Austin.

Bigelovia. DC. . . . . . . RAYLESS GOLDEN-ROD.

B. nudata, DC. Sparingly in pine barren regions. Near Blue Ball, Monmouth Co., O. R. Willis.

Chrysopsis, Nutt. .   . . . . GOLDEN ASTER.

C. falcata, Ell. Confined to pine barren regions, and not common. Near Tom's River, P. D. Knieskern, N. L. Britton; Quaker Bridge, Atsion River, W. M. Canby.

C. Mariana, Nutt. Common in the southern, and sparingly in the middle counties. Mostly confined to the Yellow Drift.

Inula, L.   .   .   . . . . ELECAMPANE.

I. Helenium, L. Common Elecampane. Sparingly escaped from gardens to roadsides in the middle and northern countries. Nat. Eu.

Pluchea, Cass. . . . . MARSH–FLEABANE.

P. camphorata, DC. Salt Marsh-fleabane. Common in salt marshes.

P. bifrons, DC. Cape May, I. C. Martindale, C. F. Parker.

Baccharis, L. .   . . . GROUNDSEL TREE.

B. halimifolia, L. Groundsel Tree. Frequent along the borders of salt marshes, and occasional in swamps beyond the flow of salt water. Near Egg Harbor City, C. F. Parker.

Polymnia, L. .   . . . LEAF–CUP.

P. Uvedalia, L. Leaf-cup. Foot of cliffs near Weehawken Ferry, 1864, T. F. Allen.

Iva, L. .   . . . MARSH ELDER.

I. frutescens, L. High-water Shrub. Common on salt marshes.

Ambrosia, Tourn.   . . .   . . RAG WEED.

A. trifida, L. Great Rag Weed. Damp places. Quite common throughout.

A. artemisiæfolia, L. Hog Weed. Rag Weed. Fields and road-sides. Very common throughout.

Xanthium, Tourn. . . . . COCKLEBUR. CLOTBUR.

X. strumarium, L. Common Cocklebur. Roadsides and waste places. Quite common.

Var. echinatum, Gray. Common along the sea-shore.

X. spinosum, L. Spiny Clotbur. Common in waste places in towns and villages. Nat. Tropical America.

Eclipta, L. . . . . ECLIPTA.

E. procumbens, Michx.; *Var.* brachypoda, Gray. Rare. Red Bank, Monmouth Co., and near Weehawhen Ferry, W. H. Leggett; banks of the Delaware at Camden, C. F. Parker; in ballast at Communipaw, A. Brown.

Heliopsis, Pers. . . . . OX–EYE.

H. lævis, Pers. Common Ox-eye. Frequent. Camden, C. F. Parker; Belleville, Essex Co., H. H. Rusby; Ocean and Monmouth Cos., P. D. Knieskern; Closter, Bergen Co., C. F. Austin.

Var. scabra, Gray. South Amboy, T. F. Allen.

Rudbeckia, L. . . . . CONE–FLOWER.

R. laciniata, L. Cone-flower. Sparingly throughout the State. Camden, C. F. Parker; rare in Ocean and Monmouth Cos., P. D. Knieskern: Belleville and Springfield, Essex Co., H. H. Rusby; Plainfield, F. Tweedy.

R. hirta, L. Yellow Daisy. Fields and pastures. Common through-out, except in the pine barrens. Nat. Western States.

Helianthus, L. . . . . SUNFLOWER.

H. annuus, L. Common Sunflower. Sparingly escaped from gar-dens into waste grounds. Adv. Tropical America.

H. angustifolius, L. Narrow-leaved S. Frequent in swamps in the pine barrens, and confined to the Yellow Drift. Sea Bright, Mon-mouth Co., A. Brown.

H. giganteus, L. Tall Sunflower. Common in swamps throughout the State.

Var. ambiguus, Gray. Cape May, C. F. Parker.

H. strumosus, L. Sparingly throughout the northern and middle counties. Closter, Bergen Co., C. F. Austin; First Mt., Essex Co., H. H. Rusby; Chatham, W. H. Leggett; Camden, C. F. Parker; near Keyport, R. W. Brown.

H. divaricatus, L. Dry fields and thickets. Quite common through-out.

H. decapetalus, L.   Frequent in the northern, and sparingly in the middle counties. Weehawken and New Durham, W. H. Leggett; Warren Co., T. C. Porter; Closter, Bergen Co., C. F. Austin; Verona, Essex Co., H. H. Rusby.

H. tuberosus, L.   Jerusalem Artichoke.   Sparingly escaped from gardens.   Camden, C. F. Parker; Closter, Bergen Co., C. F. Austin; Franklin, Essex Co., H.·H. Rusby; Bergen Point, W. H. Leggett; Keyport, R. W. Brown.

Actinomeris, Nutt. . . . .   ACTINOMERIS.

A. squarrosa, Nutt.   In Meadows, N. J., Torrey Catalogue; fields about Montclair Station, A. Brown; on west bank of Passaic River under N. Y. and G. L. R. R. bridge, W. H. Rudkin; Paterson, J. C. Hornblower.

Coreopsis, L. . . . .   TICKSEED.

C. rosea, Nutt.   Pink Tickseed.   Sparingly on the Yellow Drift. Near Hightstown, Mercer Co., O. R. Willis.

C. trichosperma, Michx.   Tickseed Sunflower.   Frequent in swamps, southern and middle counties; also, Hackensack Meadows, W. M. Wolfe.

C. discoidea, Torr. & Gray.   Camden, C. F. Parker, W. M. Canby; Budd's Lake, T. C. Porter.

C. bidentoides, Nutt.   Shore of the Delaware River at Camden, C. F. Parker.

Bidens, L. . .   . . BURR-MARIGOLD. BEGGAR TICKS.

B. frondosa, L.   Common Beggar-ticks.   Waste places.   Common throughout.

B. connata, Muhl.   Swamp Beggar-ticks.   Swamps and low grounds.   Quite common throughout.

Var. comosa, Gray.   Shore of the Delaware at Camden, C. F. Parker.

B. cernua, L.   Smaller Burr-marigold.   Sparingly in swamps, northern and middle counties.   In river dredgings at Camden, C. F. Parker; Woodside, W. H. Leggett; Franklin, Essex Co., H. H. Rusby. Eu.

B. chrysanthomoides, Michx.   Larger Burr-marigold.   Swamps. Common throughout.

B. Beckii, Torr.   Water Marigold.   Newton, Sussex Co., A. P. Garber, 1867; Swartswood Lake, T. C. Porter, 1879.

B. bipinnata, L.   Spanish Needles.   Dry soil.   Common throughout.

Helenium, L. . . . . Sneeze-weed.

H. autumnale, L. Sneeze-weed. Common in low grounds, northern and middle counties; rare on the Yellow Drift. Camden, C. F. Parker.

Galinsoga, Ruiz & Pav. . . . . Galinsoga.

G. parviflora, Cav. Becoming quite common in waste places in towns and villages. Adv. South America.

Maruta, Cass. . . . . May-weed.

M. Cotula, DC. Common May-weed. Roadsides and waste places. Common throughout. Nat. Eu.

Anthemis, L. . . . . Chamomile.

A. arvensis, L. Corn Chamomile. Fields and waste places. Becoming quite common. Adv. Eu.

A. nobilis, L. Garden Chamomile. In ballast at Camden, C. F. Parker. Adv. Eu.

Achillea, L. . . . . Yarrow.

A. Millefolium, L. Common Yarrow. Milfoil. Fields and roadsides; quite common throughout. Probably mostly naturalized from Europe. Eu.

A. Ptarmica, L. Sneezewort. In ballast at Communipaw, Addison Brown. Adv. Eu.

Leucanthemum, Tourn. . . . . . . Ox-eye Daisy.

L. vulgare, Lam. Ox-eye Daisy. White-weed. Very common throughout, in fields and meadows. Nat. Eu.

L. Parthenium, Godron. Feverfew. Sparingly escaped from gardens. New Jersey, C. F. Austin; Phillipsburg, C. F. Parker; Ballast at Communipaw, A Brown. Adv. Eu.

Matricaria, Tourn. . . . . . Wild Chamomile.

M. inodora, L. In ballast at Communipaw, Addison Brown; and Camden, C. F. Parker. Adv. Eu.

M. discoidea, DC. In ballast at Camden, C. F. Parker. Adv. Pacific Coast.

Tanacetum, L. . . . . Tansy.

T. vulgare, L. Common Tansy. Escaped from cultivation in many localities. Adv. Eu.

Artemisia, L. . . . . WORMWOOD.

A. caudata, Michx.   Sea-side Wormwood.   Sandy sea beaches and also in sandy fields at a short distance from the sea.   Not very common.   Abundant about Keyport, R. W. Brown; Sandy Hook, M. Ruger.

A. vulgaris, L.   Common Mugwort.   Sparingly introduced into waste places.   Near Closter, C. F. Austin; Montclair, Essex Co., W. M. Wolfe; in waste places and ballast at Camden, C. F. Parker; and Communipaw, Addison Brown.   Adv. Eu.

A. biennis, Willd.   Biennial Wormwood.   Occasionally found in waste places near railroads.   In ballast at Camden, C. F. Parker; abundant near the abattoir at Communipaw, A. Brown.   Adv. Western States.

Gnaphalium, L. . . . . CUDWEED.

G. decurrens, Ives.   Everlasting.   Sparingly in the northern parts of the State.   Marble Hill, Warren Co., T. C. Porter; not rare in Essex Co., H. H. Rusby; Chatham, W. H. Leggett.

G. polycephalum, Michx.   Common Everlasting.   Fields and woods.   Common throughout.

G. uliginosum, L.   Low Cudweed.   Low grounds along roadsides. Common throughout, and probably to a large extent introduced from Europe.   Eu.

G. purpureum, L.   Purplish Cudweed.   Common in the southern and sparingly in the middle counties; mostly confined to the Yellow Drift.   Chatham, W. H. Leggett; Plainfield, Frank Tweedy; Camden, C. F. Parker; Long Branch, M. Ruger.

Antennaria, Gærtn. . . . . EVERLASTING.

A. margaritacea, R. Br.   Pearly Everlasting.   Frequent in the northern and middle counties, but only sparingly on the Yellow Drift.   Dry places near the coast in Ocean and Monmouth Cos., rare, P. D. Knieskern; Keyport, R. W. Brown.

A. plantaginifolia, Hook.   Plantain-leaved Everlasting.   Dry sterile soil; common throughout.

Filago, Tourn. . . . . COTTON-ROSE.

F. Germanica, L.   Herba Impia.   Dry barren fields in Monmouth and Ocean Cos., rare, P. D. Knieskern; in ballast at Camden, I. C. Martindale.   Adv. Eu.

Erechthites, Raf. . . . . FIREWEED.

E. hieracifolia, Raf.   Fireweed.   Low grounds; common throughout.

Cacalia, L. . . . . INDIAN PLANTAIN.

C. suaveolens, L.   Indian Plantain.   Rare.   Rich fence-rows, Freehold, O. R. Willis.

C. reniformis, Muhl.   Great Indian Plantain.   Banks of the Delaware near Camden, C. F. Parker; New Jersey, Gray's Manual.

C. atriplicifolia, L.   Pale Indian Plantain.   In a meadow near Camden, C. F. Parker; the only locality known in the State.

Senecio, L. . . . . GROUNDSEL.

S. vulgaris, L.   Common Groundsel.   Waste places and ballast; becoming quite common.   Fort Lee, Bergen Co., W. H. Leggett; Snake Hill, P. V. LeRoy; Warren Co., F. Knighton; Hoboken, C. F. Austin; Communipaw, A. Brown; Camden, C. F. Parker.   Adv. Eu.

S. aureus, L.   Golden Ragwort.   Swamps and moist places.   Rather common throughout, except in the pine barrens.

Var. Balsamitæ, Gray.   Sparingly in rocky places in the northern counties.   Montclair, Essex Co., H. H. Rusby.

Carduus, Tourn. . . . . PLUMELESS THISTLE.

C. nutans, L.   Musk Thistle.   In ballast at Hoboken, A. Brown; and Camden, C. F. Parker.   Adv. Eu.

Onopordon, Vaill.   . COTTON OR SCOTCH THISTLE.

O. acanthium, L.   Cotton Thistle.   Sparingly in waste places. Near Paterson, W. H. Leggett, Warren Co., F. Knighton; in ballast at Camden, C. F. Parker; and Hoboken, I. C. Martindale.   Adv. Eu.

Centaurea, L. . . . . STAR THISTLE.

C. Cyanus, L.   Bluebottle.   Sparingly escaped from gardens, and in ballast at Camden, C. F. Parker.   Adv. Eu.

C. nigra, L.   Knapweed.   Escaped near Bloomfield, Essex Co., H. H. Rusby; in ballast at Camden, C. F. Parker; and at Communipaw. A. Brown.   Adv. Eu.

C. calcitrapa, L.   Star Thistle.   Waste places and ballast at Camden, W. M. Canby, C. F. Parker.   Adv. Eu.

Cirsium, Tourn. . . . . COMMON THISTLE.

C. lanceolatum, Scop.   Common Thistle.   Fields and roadsides; Common throughout.   Nat. Eu.

C. discolor, Spreng.   Field Thistle.   Meadows and copses.   Quite common throughout.

C. muticum, Michx.   Swamp Thistle.   Rather frequent in swamps in the northern counties, but rare elsewhere.   Princeton, O. R. Willis.

C. pumilum, Spreng. Pasture Thistle. Frequent in sandy fields, southern and middle counties, and sparingly in the northern parts of the State.

C. horridulum, Michx. Yellow Thistle. Sandy woods and fields near the coast in the southern and middle counties, often growing along the margins of salt meadows.

C. arvense, Scop. Canada Thistle. Roadsides and cultivated fields. Common in the northern and middle counties, but only sparingly in the southern parts of the State. Nat. Eu.

Lappa, Tourn. . . . . BURDOCK.

L. officinalis, All. Burdock. Waste places. Common throughout. Nat. Eu.

Lampsana, Tourn. .   . . . NIPPLE–WORT.

L. communis, L. Nipple-wort. In ballast at Communipaw, A. Brown; and Camden, C. F. Parker. Adv. Eu.

Cichorium, Tourn. . . . . . SUCCORY. CHICORY.

C. Intybus, Tourn. Common Chicory. Waste places and road-sides near towns and villages. Quite common in most sections. Nat. Eu.

Krigia, Schreb. . . . . DWARF DANDELION.

K. Virginica, Willd. Dwarf Dandelion. Common on the Yellow Drift, and sparingly throughout the rest of the State.

Cynthia, Don. . . . . CYNTHIA.

C. Virginica, Don. Virginian Cynthia. Camden, C. F. Parker; near Keyport, R. W. Brown; rather rare in Ocean and Monmouth Cos., P. D. Knieskern; Essex Co., H. H. Rusby; along Cedar Brook, Plainfield, F. Tweedy; Woodridge, Bergen Co., W. H. Rudkin; and frequent in the southern parts of the State.

Leontodon, L., Juss. . . . . . FALL DANDELION.

L. autumnale, L. Fall Dandelion. Waste places; scarce. Free-hold, O. R. Willis; in ballast at Camden, C. F. Parker. Nat. Eu.

Hieracium, Tourn. . . . .   . . HAWKWEED.

H. Canadense, Michx. Canada Hawkweed. Sparingly in the north-ern counties. Island in Lake Hopatcong, Morris Co., T. C. Porter; Closter, Bergen Co., C. F. Austin; Essex Co., H. H. Rusby.

H. scabrum, Michx. Rough Hawkweed. Dry open woods; rather common throughout.

H. Gronovii, L. Hairy Hawkweed. Rather common on the Yellow Drift, and sparingly in the middle and northern counties.

H. venosum, L. Rattlesnake-weed. Common throughout the State. *Var*. subcaulescens, Gray. Frequent.

H. paniculatum, L. Panicled Hawkweed. Open woods; rather common throughout.

Nabalus, Cass. . . . . RATTLESNAKE–ROOT.

N. albus, Hook. White Lettuce. Common in the northern and middle counties, but rare on the Yellow Drift.
Var. Serpentaria, Gray. Chatham, W. H. Leggett; Camden, C. F. Parker.

N. altissimus, Hook. Tall White Lettuce. Sparingly in the northern and central parts of the State.

N. Fraseri, DC. Lion's foot. Gall of-the-earth. Common on the Yellow Drift, and sparingly in other parts of the State. Chatham and Bergen Point, W. H. Leggett; Closter, Bergen Co., C. F. Austin.
Var. integrifolius, Gray. With the type; an occasional form. Long Hill, W. H. Leggett; Atlantic and Camden Cos., C. F. Parker.

N. virgatus, DC. Slender Rattlesnake-root. Frequent in the pine barrens, and probably confined to the Yellow Drift.

N. racemosus, Hook. Hackensack Marshes, W. H. Leggett; near Snake Hill, W. M. Wolfe; formerly grew near Closter, C. F. Austin.

Taraxacum, Haller. .   . . . . .   DANDELION.

T. Dens-leonis, Desf. Common Dandelion. Fields and roadsides ; very common throughout. Probably mostly introduced from Europe.

Lactuca, Tourn. . . . . LETTUCE

L. Canadensis, L. Wild Lettuce. Common throughout the State.
Var. integrifolia, Torr. & Gray. Rather common in the southern and middle counties.
Var. sanguinea, Torr. & Gray. Sparingly on the Yellow Drift. Atlantic City, C. F. Parker.

L. Scariola, L. Prickly Lettuce. In ballast at Communipaw, A. Brown. Adv. Eu.

Mulgedium, Cass. .   . .   FALSE OR BLUE LETTUCE.

M. acuminatum, DC. Weehawken, Torrey Catalogue; Closter, Bergen Co., C. F. Austin; First Mt., Essex Co., and Arlington, Hudson Co., W. M. Wolfe. Rare.

M. leucophæum, DC. Frequent in low grounds, in most sections of the State.

7

Sonchus, L. . . . . Sow Thistle.

S. oleraceus, L. Common Sow Thistle. Waste places; rather common throughout. Nat. Eu.

S. asper, Vill. Spiny-leaved Sow Thistle. Morris Co., C. F. Austin; New Durham, W. H. Leggett; and in ballast at Camden, C. F. Parker. Nat. Eu.

S. arvensis, L. Field Sow Thistle. Bergen Point, W. H. Leggett; Ocean and Monmouth Cos., P. D. Knieskern; waste grounds and ballast at Camden, C. F. Parker. Nat. Eu.

LOBELIACEÆ.

Lobelia, L. . . . . Lobelia.

L. cardinalis, L. Cardinal-flower. Wet ground. Common throughout.

L. syphilitica, L. Great Lobelia. Low grounds. Common in the northern, and sparingly in the middle counties; rare on the Yellow Drift. Princeton, Mercer Co., and Crosswicks, Burlington Co., O. R. Willis; Keyport, R. W. Brown.

L. puberula, Michx. Downy Lobelia. Rare, and confined to the Yellow Drift. Cape May, C. F. Parker; Freehold and Lawrenceville Landing, O. R. Willis.

L. inflata, L. Indian Tobacco. Fields and roadsides. Common all over the State.

L. spicata, Lam. Spiked Lobelia. Frequent, but not very common. Warren Co., F. Knighton; near Mattcawan, Monmouth Co., R. W. Brown; Haddonfield, C. F. Parker; Plainfield, F. Tweedy; common in Essex Co., H. H. Rusby; Hightstown, O. R. Willis; rather common near New York.

L. Nuttallii, R. & S. Nuttall's Lobelia. Common in damp places in the pine barrens, and confined to the Yellow Drift.

L. Kalmii, L. Kalm's Lobelia. Sparingly on limestone rocks in the northern counties. Sussex Co., A. P. Garber; Stockholm, Sussex Co., H. H. Rusby.

L. Canbyi, Gray. Canby's Lobelia. Sparingly in sandy swamps on the Yellow Drift. Quaker Bridge, Atlantic Co., W. M. Canby, C. E. Smith; Batestown, pine barrens, W. M. Canby; Manchester, Ocean Co., T. C. Porter.

L. Dortmanna, L. Water Lobelia. Northern shore of Green Pond, Morris Co., W. H. Rudkin. Eu.

CAMPANULACEÆ.

Campanula, Tourn. . . . . BELLFLOWER.

C. rotundifolia, L. Harebell. Rocky places. Sparingly in the northern and middle counties. Weehawken, J. S. Merriam; Palisades, C. F. Austin; Princeton, O. R. Willis.

C. aparinoides, Pursh. Marsh Bellflower. Wet meadows and swamps. Frequent throughout the State.

C. rapunculoides, L. Bellflower. Well naturalized on Long Hill, near Chatham, W. H. Leggett. Adv. Eu.

Specularia, Heis. . . . . . VENUS'S LOOKING-GLASS.

S. perfoliata, A. DC. Venus's Looking-glass. Dry soil. Common throughout.

ERICACEÆ.

Gaylussacia, H. B. K. . . . . HUCKLEBERRY.

G. dumosa, Torr. & Gray. Dwarf Huckleberry. Damp sandy soil. Common in the pine barrens, and sparingly in other parts of the State.

G. frondosa, Torr. & Gr. Dangleberry. Common on the Yellow Drift, and mostly confined to that formation. Rare in the northern counties.

G. resinosa, Torr. & Gray. Black Huckleberry. Woods and low grounds. Common throughout.

Vaccinium, L. . . . . . CRANBERRY. BLUEBERRY.

V. Oxycoccus, L. Small Cranberry. New Durham Swamp, Torrey Catalogue; Otter Pond, Closter, Bergen Co., C. F. Austin; Budd's Lake, T. C. Porter; Franklin, Essex Co., W. M. Wolfe; formerly in peat-bogs of Monmouth Co., O. R. Willis. Rare, and confined to the northern counties. Eu.

V. macrocarpon, Ait. Common Cranberry. Peat-bogs. Common in most parts of the State. Not reported from Essex Co. Scarce about Closter, Bergen Co., C. F. Austin.

V. stamineum, L. Deerberry. Dry woods. Frequent throughout the State.

V. Pennsylvanicum, Lam. Dwarf Blueberry. Closter, Bergen Co., and on the Palisades, C. F. Austin; Essex Co., H. H. Rusby; Ocean and Monmouth Cos., P. D. Knieskern. Not very common.

V. Canadense, Kalm. Canada Blueberry. Budd's Lake, Morris Co., T. C. Porter.

V. vacillans, Sol. Low Blueberry. Dry woods. Rather common throughout the State.

V. corymbosum, L. Common or Swamp Blueberry. Common in swamps throughout.

Var. atrococcum, Gray. Abundant in the Secaucus Swamp, W. H. Leggett; Camden, C. F. Parker.

Chiogenes, Salisb. . . . . . Creeping Snowberry.

C. hispidula, Torr. & Gray. Creeping Snowberry. In the cedar swamp at New Durham, Cooper in Torrey Catalogue; C. F. Austin.

Arctostaphylos, Adans. . . . . . Bearberry.

A. Uva-ursi, Spreng. Bearberry. Common in the pine barren regions. Scarce on the Palisades, C. F. Austin; Paterson Falls, Cooper in Torrey Catalogue. Eu.

Epigæa, L. . . Ground Laurel. Trailing Arbutus.

E. repens, L. Trailing Arbutus. Mayflower. Common on the Yellow Drift. Rare elsewhere.

Gaultheria, Kalm. . Aromatic Wintergreen.

G. procumbens, L. Creeping Wintergreen. Common in the pine barren regions and sparingly throughout the rest of the State.

Leucothoë, Don. . . . . Leucothoë.

L. racemosa, Gray. Racemed Leucothoë. Common in moist thickets on the Yellow Drift area, and sparingly throughout the middle and northern counties. New Durham and Secaucus, W. H. Leggett; Closter, C. F. Austin; Budd's Lake, Morris Co., T. C. Porter; Montclair and Verona, Essex Co., H. H. Rusby; Plainfield, F. Tweedy.

Cassandra, Don. . . , ... Leather-leaf.

C. calyculata, Don. Leather-leaf. Closter, Bergen Co., C. F. Austin; Budd's Lake, Morris Co., T. C. Porter; Secaucus Swamp, W. H. Leggett; and common in the pine barrens. Eu.

Andromeda, L. . . . . Andromeda.

A. polifolia, L. Peat-bogs at Budd's Lake, Morris Co., T. C. Porter; C. F. Austin. Eu.

A. Mariana, L. Stagger-bush. Closter, Bergen Co., C. F. Austin; Short Hills, Plainfield, F. Tweedy; Franklin, Essex Co., H. H. Rusby; and common on the Yellow Drift.

A. ligustrina, Muhl. Low thickets. Quite common throughout.

Clethra. L. . . . . WHITE ALDER.

C. alnifolia, L. Sweet Pepperbush. Wet woods and swamps; common throughout.

Kalmia, L. . . . . AMERICAN LAUREL.

K. latifolia, L. Calico-bush. Spoon-wood. Hillsides and thickets. Rather common throughout the State.

K. angustifolia, L. Sheep-laurel. Lambkill. Common in the middle and southern counties.

K. glauca, Ait. Pale Laurel. Budd's Lake, Morris Co., T. C. Porter.

Rhododendron, L. . . . . . ROSE-BAY. AZALEA.

R. maximum, L. Great Laurel. Damp woods along the Delaware River from Bordentown northward, O. R. Willis; West Milford, Passaic Co., W. H. Rudkin; Water Gap, W. M. Wolfe; New Durham and Secaucus Swamps, and Great Swamp near Chatham, W. H. Leggett.

R. viscosum, Torr. (Azalea, L.) White Swamp Honeysuckle. Swamps. Common in the southern and middle counties, and sparingly in the northern parts of the State. Franklin, Essex Co., H. H. Rusby.

Var. nitidum, Gray. Franklin, Essex Co., H. H. Rusby.

R. nudiflorum, Torr. (Azalea, L.) Pinxter-flower. Damp woods. Common throughout.

Rhodora, Duham. . . . . RHODORA.

R. Canadensis, L. In a bog at Succasunna, Morris Co., T. C. Porter.

Leiophyllum, Pers. . . . . . . SAND MYRTLE.

L. buxifolium, Ell. Sand Myrtle. Common in the pine barrens, and confined to the Yellow Drift.

Pyrola, Tourn. . . . . WINTERGREEN. SHIN-LEAF.

P. rotundifolia, L. Common in the northern and frequent in the middle counties; rare on the Yellow Drift. Ocean and Monmouth Cos., P. D. Knieskern; near Keyport, R. W. Brown. Eu.

P. elliptica, Nutt. Shin-leaf. Common, except on the Yellow Drift.

P. chlorantha, Swartz. Rare. Dry woods, near Camden, C. F. Parker; Warren Co., F. Knighton; near Closter, C. F. Austin, A. Brown.

P. secunda, L. Freehold, O. R. Willis; Closter, Bergen Co., C. F. Austin; Sussex Co., A. P. Garber; Camden Co., C. F. Parker; open sandy woods, N. J., Torrey Catalogue. Eu.

Chimaphila, Pursh. . . . . . . Pipsissewa.

C. umbellata, Nutt. Prince's Pine. Dry woods. Common throughout the State. Eu.

C. maculata, Pursh. Spotted Wintergreen. Dry woods. Common
throughout.

Monotropa, L. . . . . . Indian Pipe. Pine-sap.

M. uniflora, L. Indian Pipe. Dark rich woods. Quite common
throughout.

M. Hypopitys, L. Pine-sap. Oak or pine woods. Sparingly
throughout the State. Eu.

EBENACEÆ.

Diospyros, L. . . . . . Date plum. Persimmon.

D. Virginiana, L. Common Persimmon. Old fields and borders of
woods. Common in the southern and middle counties and sparingly
in the northern part of the State. Chatham and Bergen Point, W.
H. Leggett; White House Station, C. R. R. of N. J., C. F. Austin;
Franklin, Essex Co., H. H. Rusby.

DIAPENSIACEÆ.

Pyxidanthera, Michx. . . . . . Pyxidanthera.

P. barbulata, Michx. Flowering Moss. Common in the pine barrens and confined to the Yellow Drift.

PLUMBAGINACEÆ.

Statice, Tourn. . . . . Marsh-rosemary.

S. Limonium, L.; *Var.* Caroliniana, Gray. Common on salt
marshes. Eu.

PRIMULACEÆ.

Trientalis, L. . . . . . Chickweed-wintergreen.

T. Americana. Pursh. Star-flower. Damp woods. Common in
the northern and middle counties, but grows only sparingly on the
Yellow Drift. Near Keyport, Monmouth Co., R. W Brown; Ocean
and Monmouth Cos., P. D. Knieskern; Atlantic City, C. F. Parker.

Lysimachia, Tourn. . . . . Loosestrife.

L. thrysiflora, L. Tufted Loosestrife. Hackensack marshes, W.
H. Leggett; Bergen Co., C. F. Austin; west shore of Swartswood Lake,

Sussex Co., W. H. Rudkin; Budd's Lake, Morris Co., T. C. Porter.
Rare, and confined to the northern counties. Eu.

L. stricta, Ait. Racemed Loosestrife. Low grounds. Common
throughout.

L. quadrifolia, L. Whorled Loosestrife. Low grounds. Common
throughout.

L. nummularia, L. Moneywort. Sparingly escaped from cultivation. Montclair, Essex Co., W. M. Wolfe. Adv. Eu.

Steironema, Raf. (Lysimachia, Tourn.) . LOOSESTRIFE.

S. ciliatum, L. Ciliate-leaved Loosestrife. Low grounds and
thickets. Quite common throughout.

S. lanceolatum, Gray. Narrow-leaved Loosestrife. Low grounds.
Quite common in the middle and northern counties.

Var. hybridum, Gray. Newark Meadows, Torrey Catalogue; Franklin, Essex Co., H. H. Rusby.

Var. angustifolium, Gray. Franklin, Essex Co., H. H. Rusby.

Anagallis, Tourn. . . . . PIMPERNEL.

A. arvensis, L. Poor Man's Weather-glass. Waste fields and roadsides. Quite common in most places. Nat. Eu.

Samolus, L. . . WATER PIMPERNEL. BROOK-WEED.

S. Valerandi, L.; *Var.* Americanus, Gray. Frequent along the borders of salt marshes. The typical S. Valerandi, L., introduced in ballast
at Camden, C. F. Parker.

Hottonia, L. . . . . WATER VIOLET.

H. inflata, Ell. Featherfoil. Closter, Bergen Co., C. F. Austin;
Fort Lee, Palisades, W. H. Leggett; Eatontown, Monmouth Co., O.
R. Willis; on the Palisades near Guttenberg and Pleasant Valley, W.
H. Rudkin; Carrieville Station, N. R. R. of N. J., Dr. John Torrey;
one mile east of Woodridge Station, Hackensack Branch N. Y., L. E.
and W. R. R., G. C. Woolson.

OLEACEÆ.

Ligustrum, Tourn. . . . . PRIVET.

L. vulgare, L. Common Privet or Prim. Roadsides and copses;
occasional. Ocean and Monmouth Cos., P. D. Knieskern; Keyport,
R. W. Brown; Stanhope, Sussex Co., C. F. Austin. Nat. Eu.

Fraxinus, Tourn. . . . . ASH.

F. Americana, L. White Ash. Moist woods. Common in the
northern and middle counties.

F. pubescens, Lam. Red Ash. Moist woods, with the same range as the last species, but less common.

F. sambucifolia, Lam. Black Ash. Closter, C. F. Austin; common in Essex Co., H. H. Rusby; Freehold, O. R. Willis. Rare, and mostly confined to the northern counties.

APOCYNACEÆ.

Vinca, L. . . . . Periwinkle.

V. minor, L. Creeping Periwinkle. Sparingly escaped from cultivation in many localities. Adv. Eu.

Apocynum, Tourn. . . . Dogbane. Indian Hemp.

A. androsæmifolium, L. Spreading Dogbane. Fields and borders of thickets. Sparingly in the southern and middle counties, but common in the northern parts of the State.

A. cannabinum, L. Indian Hemp. Low grounds and banks of streams. Common throughout.

Var. pubescens, DC. Frequently found with the type.

ASCLEPIADACEÆ.

Asclepias, L. . . . . . . Milkweed. Silkweed.

A. Cornuti, Dec. Common Milkweed. Fields and roadsides. Common throughout the State.

A. phytolaccoides, Pursh. Poke Milkweed. Moist copses. Occasional in the northern and middle counties. Nowhere very abundant.

A. purpurascens, L. Purple Milkweed. Frequent in the northern and middle counties; rare on the Yellow Drift.

A. variegata, L. Variegated Milkweed. Rare. Closter, Bergen Co., C. F. Austin; near Smithville, O. R. Willis; near Camden, 1863, (locality destroyed) C. F. Parker.

A. quadrifolia, Jacq. Four-leaved Milkweed. Dry woods. Frequent in the northern and middle counties.

A. incarnata, L. Swamp Milkweed. Frequent in low grounds, middle and northern counties.

Var. pulchra, Gray. Swamp Milkweed. Low grounds. Common throughout.

A. obtusifolia, Michx. Wavy-leaved M. Common on the Yellow Drift, and sparingly elsewhere. Franklin, Essex Co., H. H. Rusby.

A. rubra, L. Red Milkweed. Sparingly in pine barren regions.

A. paupercula, Michx. Sparingly in pine barren regions.

A. tuberosa, L. Butterfly-weed. Dry hills and fields. Common in the middle and southern counties, and frequent in the northern parts of the State.

A. verticillata, L. Whorled Milkweed. Palisades, C. F. Austin; on the rocks at Passaic Falls, Torrey Catalogue; Red Bank, Monmouth Co., W. H. Leggett; Freehold, O. R. Willis; Montclair Heights, Essex Co., W. H. Rudkin; First Mt. near Verona, Essex Co., H. H. Rusby; near Holmdel, Monmouth Co., S. Lockwood; Clarksboro, Gloucester Co., I. Burk.

Acerates, Ell.   . . . . . .   GREEN MILKWEED.

A. viridiflora, Ell. Green Milkweed. Dry pine woods, Monmouth Co., O. R. Willis; Black's Mills, S. Lockwood; Closter, Bergen Co., 1858, C. F. Austin; and probably frequent in the southern counties.

LOGANIACEÆ.

Polypremum, L.   . . . .   POLYPREMUM.

P. procumbens, L. In ballast at Camden, C. F. Parker. Adv. Southern States.

GENTIANACEÆ.

Sabbatia, Adans.   . . . . .   AMERICAN CENTAURY.

S. lanceolata, Torr. & Gray. Frequent in the pine barrens and confined to the Yellow Drift.

S. angularis, Pursh. Not common. Closter, Bergen Co., C. F. Austin; foot of hill opposite Montclair Heights Station, W. H. Rudkin; Franklin, Essex Co., H. H. Rusby; Long Hill and Stony Hill, W. H. Leggett; New Durham, P. V. Leroy; near Keyport, Monmouth Co., R. W. Brown; Cape May, C. F. Parker.

S. stellaris, Pursh. Common along the borders of salt marshes.

S. chloroides, Pursh. Hackensack marshes, W. H. Leggett; banks of Mullica River, near Pleasant Mills, Ocean Co., and in marshes at Cape May, C. F. Parker.

Erythræa, Pers.   . . .   .   CENTAURY.

E. ramosissima, Pers., *Var.* pulchella, Griseb. Closter, Bergen Co., C. F. Austin, 1858. Nat. Eu.

Gentiana, L.   .   . . .   GENTIAN.

G. quinqueflora, Lam. Five-flowered G. In woods about Elizabethtown, Eddy in Torrey Catalogue; Sussex Co., C. F. Austin; Free-

hold and Hightstown, O. R. Willis ; moist hills, Morris Co., T. C. Porter; Warren Co., A. P. Garber. Rare, and mostly confined to the northern counties.

G. crinita, Frœl. Fringed Gentian. Frequent. Totowa Mts., N. J., Torrey Catalogue ; common about Closter, C. F. Austin ; Hackensack marshes and Long Hill, W. H. Leggett ; Hightstown and Freehold, O. R. Willis ; Roseland, Essex Co., H. H. Rusby ; common about Plainfield, Frank Tweedy ; N. R. R. between Tenafly and Cresskill, W. H. Rudkin ; near Keyport, Monmouth Co., R. W. Brown ; Morris Co., T. C. Porter ; Warren Co., A. P. Garber.

G. Andrewsii, Griseb. Closed Gentian. Shark River, Ocean Co., P. D. Knieskern ; near Keyport, Monmouth Co., R. W. Brown ; frequent in the middle and common in the northern counties.

G. Saponaria, L. Soapwort Gentian. Cresskill, N. R. R. of N. J., I. H. Hall ; Plainfield, Frank Tweedy ; Shark River, Ocean Co., P. D. Knieskern ; and frequent in the southern counties.

Var. linearis, Gray. Budd's Lake, Morris Co., T. C. Porter.

G. angustifolia, Michx. Pine-barren G. Sparingly in pine barren regions.

Bartonia, Muhl.   .   .   .   . BARTONIA.

B. tenella, Muhl. Moist open woods. Frequent throughout the State.

Obolaria, L. .   .   .   .   ,   .   .   . OBOLARIA.

O. Virginica, L. Virginian Obolella. Moist woods near Bloomsbury, Sussex Co., A. P. Garber ; near Princeton and about Lawrenceville, O. R. Willis ; Montclair, Essex Co., H. H. Rusby. Rare and local.

Menyanthes, Tourn. .   .   .   .   .   .   BUCKBEAN.

M. trifoliata, L. Three-leaved Buckbean. In ponds. Rare and confined to the northern parts of the State. One mile south-east of Closter, Bergen Co., C. F. Austin ; N. R. R. of N. J., near its junction with N. Y. L. E. & W. R. R., W. H. Leggett ; New Durham, P. V. LeRoy ; Budd's Lake, Morris Co., T. C. Porter ; bogs, Sussex Co., A. P. Garber. Eu.

Limnanthemum, Gmelin.   .   . FLOATING HEART.

L. lacunosum, Griseb. In ponds, New Jersey, Eddy in Torrey Catalogue. Sparingly in pine barren regions.

POLEMONIACEÆ.

Polemonium, Tourn. .   .   .   .   GREEK VALERIAN.

P. cæruleum, L. Jacob's Ladder. In a swamp near Washington, Warren Co., A. P. Garber ; C. F. Parker ; T. C. Porter. Eu.

Phlox, L. . . . . PHLOX.

P. paniculata, L. Panicled Phlox. Sparingly escaped from cultivation to roadsides and waste places. Adv. Western States.

P. maculata, L. Wild Sweet William. Sparingly escaped from
cultivation to roadsides and waste places. Adv. Western States.

P. pilosa, L. Passaic, Dr. George Thurber ; Chatham, W. H. Leg-
gett; Plainfield, Frank Tweedy; Milburn, Essex Co., N. L. Britton;
Woodbury, Gloucester Co., W. M. Canby ; near New Brunswick, S.
Lockwood. Not common.

P. subulata, L. Ground or Moss Pink. Occasional in the middle
and northern counties. Near Paterson, C. F. Austin ; Warren and
Hunterdon Cos., F. Knighton ; Great Notch, Passaic Co., W. M.
Wolfe; sandy fields, Union Co., F. Tweedy; Camden Co., C. F.
Parker; Red Bank, Monmouth Co., W. H. Leggett; New Brunswick,
S. Lockwood ; hill opposite Montclair Heights Station, Essex Co., W.
H. Rudkin.

HYDROPHYLLACEÆ.

Hydrophyllum, L. .   . . . WATERLEAF.

H. Virginicum, L. Common Waterleaf. New Durham Swamp, W.
H. Leggett; Weehawken, W. H. Leggett, N. L. Britton ; Freehold, O.
R. Willis; Little Falls, Passaic Co., W. M. Wolfe; Plainfield, F. Tweedy;
Essex Co., H. H. Rusby. Not common, and mostly confined to the
northern counties.

Ellisia, L. . . .   . ELLISIA.

E. Nyctelea, L. Banks of the Delaware River, near Trenton, W.
M. Canby.

BORRAGINACEÆ.

Echium, Tourn. . . . . . .   VIPER'S BUGLOSS.

E. vulgare, L. Blue-weed. Sparingly in fields and along roadsides
throughout the State. Near Keyport, R. W. Brown ; Monmouth and
Ocean Cos., P. D. Knieskern; Bergen Point, W. H. Leggett; Wee-
hawken, C. F. Austin; George's Road, near cross roads, Middlesex
Co, O. R. Willis; Warren Co., F. Knighton ; Franklin, Essex Co., H.
H. Rusby ; New Brunswick, and all along the P. R. R., from Jersey
City westward, N. L. Britton ; Delaware Water Gap and in ballast at
Camden, C. F. Parker; Plainfield, F. Tweedy ; in ballast at Commu-
nipaw, A. Brown. Nat. Eu.

Lycopsis, L. . . . . BUGLOSS.

L. arvensis, L. Small Bugloss. Bergen Co., W. H. Leggett; in ballast at Camden, C. F. Parker. Adv. Eu.

Symphytum, Tourn.   . . . . COMFREY.

S. officinale, L. Common Comfrey. Sparingly escaped from cultivation to roadsides and waste places. Plainfield, Frank Tweedy; Verona, Essex Co., H. H. Rusby; in ballast at Communipaw and Hoboken, A. Brown. Adv. Eu.

Onosmodium, Michx. . . . . . FALSE GROMWELL.

O. Virginianum, DC. Rare. Camden, C. F. Parker; south side of Chesquake Creek, Middlesex Co., and Brown's Point, Keyport, Monmouth Co., R. W. Brown.

Mertensia, Roth. . . . . . . SMOOTH LUNGWORT.

M. Virginica, DC. Virginian Cowslip. Near Walnford, Monmouth Co., S. Lockwood. See Bull. Torr. Bot. Club, May, 1881.

Lithospermum, Tourn.   . GROMWELL. PUCCOON.

L. arvense, L. Corn Gromwell. Sandy banks. Ocean and Monmouth Cos., rare, P. D. Knieskern; waste grounds and ballast at Camden, C. F. Parker; ballast at Communipaw, A. Brown; and sparingly in waste places along the railroads. Nat. Eu.

L. officinale, L. Common Gromwell. Plentiful at the New Jersey Zinc Mines, Sussex Co., C. F. Austin; hills back of Hoboken, M. Ruger. Nat. Eu.

Myosotis, L. . . . .   FORGET-ME-NOT.

M. palustris, Withering; *Var.* laxa, Gray. Wet places. Common in the northern and middle counties. Keyport, Monmouth Co., R. W. Brown; Camden, C. F. Parker. Probably mostly escaped from cultivation. Eu.

M. arvensis, L. Closter, Bergen Co., C. F. Austin in Willis Catalogue. Eu.

M. verna, Nutt. Scorpion-grass. Dry hills. Frequent throughout the State.

Echinospermum, Swartz. . . . . . . STICKSEED.

E. Lappula, Lehm. Stickseed. Rare. Hoboken, and in ballast at Communipaw, Addison Brown; in ballast at Camden, C. F. Parker; Freehold, O. R. Willis. Nat. Eu.

Cynoglossum, Tourn. . . . . . . Hound's-tongue.

C. officinale, L.   Common Hound's-tongue.   Waste places.   Not very common.   Freehold, S. Lockwood; Essex Co., H. H. Rusby; Palisades, C. F. Austin; Warren Co., F. Knighton; Weehawken, N. L. Britton.   Nat. Eu.

C. Virginicum, L.   Wild Comfrey.   Rare.   First Mt., Essex Co., R. Spaulding; near Plainfield, F. Tweedy; Hunterdon Co., C. F. Parker; Weehawken, and Fort Lee, Bergen Co., W. H. Leggett.

C. Morrisoni, DC.   Beggar's Lice.   Woods and copses.   Sparingly throughout the State.

Heliotropium, Tourn. . . . . Heliotrope.

H. Europæum, L.   European Heliotrope.   In ballast at Communipaw, A. Brown; and at Camden, C. F. Parker.   Adv. Eu.

CONVOLVULACEÆ.

Ipomœa, L. . . . . Morning Glory.

I. coccinea, Mœnch.   (Quamoclit, Tourn.)   Scarlet Morning Glory. In ballast at Communipaw, A. Brown; and at Camden, C. F. Parker. Adv. Tropical America.

I. purpurea, Lam.   Common Morning Glory.   Commonly escaped from gardens into waste places.   Also in ballast.   Adv. Tropical America.

I. Nil. Roth.   Smaller Morning Glory.   Bank of the Passaic River near Belleville, N. L. Britton, 1879; near Union, Monmouth Co., R. W. Brown; in ballast at Camden, C. F. Parker.   Adv. Tropical America. (?)

I. pandurata, Meyer.   Wild Potato-vine.   Closter, Bergen Co., C. F. Austin; Warren Co., F. Knighton; New Providence, W. H. Leggett; and frequent in the southern and middle counties.

Convolvulus, L. . . . . Bindweed.

C. arvensis, L.   Bindweed.   Roadsides near Closter, rare, C. F. Austin; Newark Meadows along C. R. R. of N. J., W. M. Wolfe; Freehold, S. Lockwood; Newton, Sussex Co., N. L. Britton; in ballast at Camden, C. F. Parker; and at Communipaw, A. Brown.   Nat. Eu.

C. sepium, L.   (Calystegia, R. Br.)   Hedge Bindweed.   Low grounds along streams and along fences and hedges.   Common throughout.

C. spithamæus, L.   (Calystegia, R. Br.)   Downy Bindweed.   Probably grows within the State but is certainly rare and no definite localities are reported.

Breweria, R. Br. (Bonamia, Thouars.) . . Breweria.

B. Pickeringii, Gray. Sparingly in dry sandy pine barrens in the southern parts of the State.

Cuscuta, Tourn. . . . . Dodder.

C. Epilinum, Weihe. Flax Dodder. Flax fields; not common. P. D. Knieskern in Catalogue of Plants of Monmouth and Ocean Cos. Adv. Eu.

C. tenuiflora, Engelm. Quaker Bridge, Atlantic Co., Dr. George Engelmann. Rare.

C. arvensis, Beyrich. Sandy fields, Cape May, C. F. Parker; Closter, Bergen Co., C. F. Austin; Shark River, Ocean Co., Forman in Bull. Torr. Bot. Club, vol. 2, p. 36.

C. Gronovii, Willd. Low grounds. Common throughout.

C. compacta, Juss. Common in Bergen and Ocean Cos., C. F. Austin; pine barrens, W. M. Canby; Keyport, Monmouth Co., R. W. Brown; Camden, C. F. Parker.

SOLANACEÆ.

Solanum, Tourn. . . . . Nightshade.

S. Dulcamara, L. True Bittersweet. Frequent in waste places near houses. Nat. Eu.

S. nigrum, L. Common Nightshade. Waste places. Rather common all over the State. Nat. Eu.

S. Carolinense, L. Horse-nettle. Not common. Phillipsburg, T. C. Porter; Camden, C. F. Parker; Bergen Point, W. H. Leggett; near Weehawken, N. L. Britton; banks of Delaware River near Phillipsburg, T. C. Porter.

Physalis, L. . . . . Ground Cherry.

P. Philadelphica, Lam. Springfield and Franklin, Essex Co., H. H. Rusby; and probably elsewhere in the State.

P. pubescens, L. Rare. Closter, Bergen Co., C. F. Austin.

P. viscosa, L. Sandy fields. Common throughout the State.

Nicandra, Adans. . . . . Apple of Peru.

N. physaloides, Gærtn. Waste grounds and ballast at Camden, C. F. Parker; Ocean Grove, and in ballast at Communipaw, A. Brown. Adv. Peru.

Lycium, L.   . . . . Matrimony–vine.

L. vulgare, Dunal. Matrimony-vine. Sparingly escaped from gardens to roadsides. Adv. Eu.

Hyoscyamus, Tourn. . . . . HENBANE.

H. niger, L. Black Henbane. In ballast at Communipaw, A. Brown; and at Camden, Isaac Burk; Warren Co., F. Knighton. Adv. Eu.

Datura, L. . . . JAMESTOWN WEED. THORN APPLE.

D. Stramonium, L. Common Thorn Apple. Waste places and cultivated fields. Common throughout. Adv. Asia.

D. Tatula, L. Purple Thorn Apple. Phillipsburg, T. C. Porter; Newark, H. H. Rusby; near Keyport, Monmouth Co., R. W. Brown; frequent in waste places near New York, N. L. Britton. Adv. Tropical America. (?)

SCROPHULARIACEÆ.

Verbascum, L. . . . . MULLEIN.

V. Thapsus, L. Common Mullein. Old fields and roadsides. Common throughout. Nat. Eu.

V. Blattaria, L. Moth Mullein. Fields and waste places. Common throughout. Nat. Eu.

V. Lychnitis, L. White Mullein. Waste places. Rare. Warren Co., F. Knighton; Pavonia, near Camden, C. F. Parker; Trenton, S. Lockwood. Adv. Eu.

Linaria, Tourn. .   . . . .   . TOAD-FLAX.

L. Canadensis, Spreng. Wild Toad-flax. Sandy soil. Common throughout.

L. vulgaris, Mill. Butter and Eggs. Old fields and roadsides. Common throughout. Nat. Eu.

L. Elatine, Mill. In ballast at Communipaw, A. Brown; and at Camden, C. F. Parker. Adv. Eu.

Antirrhinum, L. . . . . SNAPDRAGON.

A. Orontium, L. In ballast at Communipaw, A. Brown; and at Camden, C. F. Parker. Adv. Eu.

Scrophularia, Tourn. . . . . FIGWORT.

S. nodosa, L. Figwort. Frequent throughout the State. Camden, C. F. Parker; Essex Co., H. H. Rusby; Ocean and Monmouth Cos., P. D. Knieskern; rare at Closter, C. F. Austin; Palisades, N. L. Britton.

Chelone, Tourn. . . . TURTLE-HEAD. SNAKE-HEAD.

C. glabra, L. Shell-flower. Wet places. Rather common throughout.

Pentstemon, Mitchell. . . . . . . BEARD-TONGUE.

P. pubescens, Sol. Beard-tongue. Not common. Closter, Bergen Co., C. F. Austin; Preakness, W. L. Fischer; abundant near Pompton, and at Bergen Point, W. H. Leggett; Princeton and Freehold, O. R. Willis; Montclair, Essex Co., W. M. Wolfe; abundant near Plainfield, F. Tweedy; common along the Delaware, T. C. Porter.

Mimulus, L. . . . . MONKEY-FLOWER.

M. ringens, L. Common Monkey-flower. Low grounds. Common in the northern and middle counties, and sparingly on the Yellow Drift.

M. alatus, Ait. Winged Monkey-flower. Low grounds. Rare. Camden, C. F. Parker; Ogdensburg, Sussex Co., and Bloomfield, Essex Co., H. H. Rusby; Monmouth and Ocean Cos., P. D. Knieskern.

Conobea, Aublet. . . . . CONOBEA.

C. multifida, Benth. In ballast at Camden, C. F. Parker. Adv. Western States.

Herpestis, Gærtn. . . . . HERPESTIS.

H. amplexicaulis, Pursh. New Jersey, Gray's Manual, p. 329.

Gratiola, L. . . . . HEDGE-HYSSOP.

G. Virginiana, L. Wet places. Common throughout the State.

G. sphærocarpa, Ell. New Jersey, Gray's Manual, p. 330; Cape May, C. F. Austin.

G. aurea, Muhl. Lake Hopatcong, H. H. Rusby; shore of Delaware River above Phillipsburg, T. C. Porter; and common on the Yellow Drift.

G. pilosa, Mich. Low ground near Camden, C. E. Smith, W. M. Canby; Cape May, C. F. Parker.

Ilysanthes, Raf. . . . . ILYSANTHES.

I. gratioloides, Benth. False Pimpernel. Wet places. Rather common throughout.

Micranthemum, Michx. . . . MICRANTHEMUM.

M. Nuttallii, Gray. Tidal mud, banks of the Delaware at Camden, C. F. Parker.

Limnosella, L. . . . . MUDWORT.

L. aquatica, L; *Var.* tenuifolia, Hoffm. Long Branch, Torrey Catalogue; Passaic River. W. H. Leggett; Hackensack River, Bergen Co., C. F. Austin. Rare.

Veronica, L. . . . . SPEEDWELL.

V. Virginica, L. Culver's-root. Not common. Bergen Co., C. F. Austin; Plainfield, F. Tweedy; occasional in Essex Co., H. H. Rusby; Chatham, W. H. Leggett; Monmouth Co., O. R. Willis; near Chesquakes Creek, Middlesex Co., R. W. Brown.

V. Anagallis, L. Water Speedwell. Sparingly in the northern counties. Bergen Co., C. F. Austin; N. R. R. of N. J., W. H. Leggett; Morris Co., T. C. Porter; Franklin, Essex Co., H. H. Rusby; also Hightstown, O. R. Willis. Eu.

V. Americana, Schwein. American Brooklime. Frequent in the northern and middle couuties. Rare on the Yellow Drift. New Egypt, Ocean Co., P. D. Knieskern.

V. scutellata, L. Marsh Speedwell. Frequent in the northern and middle counties. Rare on the Yellow Drift. Eu.

V. officinalis, L. Common Speedwell. Woodlands and roadsides. Quite common throughout. Probably indigenous in the northern counties, but introduced elsewhere. Eu.

V. serpyllifolia, L. Thyme-leaved Speedwell. Fields and roadsides. Common throughout. Certainly introduced in part. Eu.

V. peregrina, L. Purslane Speedwell. Waste and cultivated grounds. Common throughout the State.

V. arvensis, L. Corn Speedwell. Roadsides and cultivated grounds. Rather common throughout. Nat. Eu.

V. agrestis, L. Field Speedwell. In ballast at Camden, C. F. Parker; and Communipaw, A. Brown. Adv. Eu.

V. Buxbaumii, Tenore. Buxbaum's Speedwell. In ballast at Camden, C. F. Parker; and Communipaw, N. L. Britton. Adv. Eu.

V. hederæfolia, L. Ivy-leaved Speedwell. Weehawken, M. Ruger; ballast at Camden, C. F. Parker; and at Hoboken and Communipaw, A. Brown. Adv. Eu.

Gerardia, L . . . . GERARDIA.

G. purpurea, L. Purple Gerardia. Low grounds. Common except in the northern parts of the State. A form with white flowers at Atlantic City, C. F. Parker.

G. maritima, Raf. Sea-side Gerardia. Frequent on salt meadows.

G. tenuifolia, Vahl. Slender Gerardia. Dry woods Common throughout.

G. flava, L. Downy False Foxglove. Open woods. Quite common throughout.

G. quercifolia, Pursh. Smooth False Foxglove. Rare. Morristown, W. H. Leggett; Freehold, S. Lockwood; Closter, Bergen Co., C. F. Austin; Stockholm, Sussex Co., H. H. Rusby.

G. pedicularia, L. Lousewort Gerardia. Hohokus, C. F. Austin; Stockholm, Sussex Co., H. H. Rusby; and common in the southern and middle counties.

Castilleia, Mutis. . . . .   PAINTED-CUP.

C. coccinea, Spreng. Scarlet Painted-cup. Chatham, W. H. Leggett; Closter, Bergen Co., C. F. Austin; Feltville, F. Tweedy; sparingly in Monmouth and Mercer Cos., O. R. Willis; Roseland and Northfield, Essex Co., H. H. Rusby; borders of pond four miles west of Newton, Sussex Co., Arthur Hollick; Clifton, Passaic Co., W. H. Rudkin; West Orange, W. M. Wolfe.

Schwalbea, Gronov. . . . .   CHAFF SEED.

S. Americana, L. Sparingly in pine barren regions. Abundant near Egg Harbor City, C. F. Parker.

Pedicularis, Tourn. . . . .   LOUSEWORT.

P. Canadensis, L. Common Lousewort. Copses and banks. Common throughout.

P. lanceolata, Michx. Hackensack Meadows, Torrey Catalogue; Chatham, W. H. Leggett; Freehold, O. R. Willis; Verona, Essex Co., H. H. Rusby; common at Closter, C. F. Austin; Newton, Sussex Co., A. P. Garber; Plainfield, F. Tweedy.

Melampyrum, Tourn. . . . . COW-WHEAT.

M. Americanum, Michx. Open woods. Common throughout.

OROBANCHACEÆ.

Epiphegus, Nutt. . . . . CANCER-ROOT.

E. Virginiana, Bart. Beech-drops. Parasitic on the roots of beech-trees. Common in the northern and middle counties.

Conopholis, Wallroth. . . . . . CANCER-ROOT.

C. Americana, Wallroth. Squaw-root. Parasitic on the roots of oaks and other trees. Frequent in the northern and middle counties.

Orobanche, L. . . . . BROOM-RAPE.

O. minor, L. Small Broom-rape. Parasitic on clover near Camden, C. F. Parker; Haddonfield, J. H. Redfield. Adv. Eu

Aphyllon, Mitchell. . . . . . NAKED BROOM-RAPE.

A. uniflorum, T. & G. One-flowered Cancer-root. Generally and perhaps always parasitic on Solidagos. (See Prof. Jos. Schrenck, in Bull. Torr. Bot. Club, vol. vii., p. 67.) Frequent throughout the State.

Ocean and Monmouth Cos., P. D. Knieskern; Plainfield, F. Tweedy; common in Essex Co., H. H. Rusby; most abundant in the northern counties.

LENTIBULACEÆ.

Utricularia, L.   . . . . BLADDERWORT.

U. inflata, Walt.   Inflated Bladderwort.   Ponds.   Not common. Stagnant waters in New Jersey, Eddy in Torrey Catalogue; Long Branch, O. R. Willis; Egg Harbor City and Woodbury, C. F. Parker; Plainfield, Frank Tweedy.   Mostly confined to the Yellow Drift.

U. vulgaris, L.; *Var.* Americana, Gray.   Ditches and slow streams. Rather common throughout.

U. clandestina, Nutt.   Sparingly in pine barren regions.   Monmouth Co., O. R. Willis; Atsion, W. M. Canby.

U. intermedia, Hayne.   Rare.   Budd's Lake, T. C. Porter; Closter, Bergen Co., C. F. Austin; near Camden, C. F. Parker.   Eu.

U. fibrosa, Walt.   (**U.** striata, Le Conte.)   Sparingly on the Yellow Drift.   Camden Co., Burlington Co., Quaker Bridge, Atlantic Co., Elmer, Salem Co., C. F. Parker; Upper Squankum, Monmouth Co., O. R. Willis.

U. gibba, L.   Closter, Bergen Co., C. F. Austin; near the Passaic at Woodside, W. H. Leggett.   Rare.

U. purpurea, Walt.   Sparingly in ponds in the pine barrens.

U. cornuta, Michx.   Sandy borders of ponds.   Common on the Yellow Drift and sparingly in other parts of the State.

U. subulata, L.   Frequent in sandy swamps in pine barren regions.

BIGNONIACEÆ.

Tecoma, Juss. . . . . TRUMPET-FLOWER.

T. radicans, Juss.   Trumpet Creeper.   Frequently escaped from cultivation.

Catalpa, Scop., Walt. . . . .   . . CATALPA.

C. bignonioides, Walt.   Indian Bean.   Extensively planted as an ornamental tree, and sometimes escaping from cultivation.

VERBENACEÆ.

Verbena, L. . . . . VERVAIN.

V. angustifolia, Michx.   Narrow-leaved V.   Sparingly throughout the State.   Hoboken, Torrey Catalogue; Passaic Falls, J. S. Merriam; Montclair, Essex Co., W. M. Wolfe; Closter, Bergen Co., C. F. Austin;

Red Bank, W. H. Leggett; Long Branch, M. Ruger; Seabright, N. L. Britton; Plainfield, F. Tweedy; Jackson, Camden Co., C. F. Parker; Atco, Camden Co., I. H. Hall.

V. hastata, L. Blue Vervain. Low grounds and roadsides. Common throughout.

V. urticæfolia, L. White Vervian. Roadsides and waste places. Common throughout.

V. officinalis, L. European Vervain. Roadsides and ballast at Camden, C. F. Parker; ballast at Communipaw, A. Brown. Adv. Eu.

V. bracteosa, Michx. In ballast at Camden, C. F. Parker. Adv. Western States.

Phryma, L. . . . . Lopseed.

P. Leptostachya, L. Lopseed. Woods and copses. Quite common in all parts of the State.

LABIATÆ.

Teucrium, L. . . . . Germander.

T. Canadense, L. American Germander. Low grounds. Frequent throughout.

Trichostema, L. . . . . Blue Curls.

T. dichotomum, L. Bastard Pennyroyal. Dry fields. Rather common throughout.

T. lineare, Nutt. Sparingly in the pine barrens and confined to the Yellow Drift formation.

Isanthus, Michx. . . . . False Pennyroyal.

I. caeruleus, Michx. Rare. Freehold, O. R. Willis; Bergen Co., C. F. Austin.

Mentha, L. . . . . Mint.

M. rotundifolia, L. Round-leaved Mint. Hunterdon Co., on the Delaware, T. C. Porter; Bloomfield, and Hudson Station, N. R. R. of N. J., W. H. Leggett. Adv. Eu.

M. viridis, L. Spearmint. Wet places. Common throughout the State. Nat. Eu.

M. piperita, L. Peppermint. Low grounds and along brooks. Quite common throughout. Nat. Eu.

M. aquatica, L. Water Mint. Wet ballast, Camden, C. F. Parker. Adv. Eu.

Var. crispa, Benth. Pamrapo, on Bergen Neck, W. H. Leggett. Adv. Eu.

M. sativa, L.; *Var.* glabra. Whorled Mint. River banks, Phillips-burg, T. C. Porter. Adv. Eu.

M. arvensis, L. Corn Mint. In ballast at Camden, C. F. Parker. Adv. Eu.

M. Canadensis, L. Wild Mint. Damp places. Frequent in the middle and northern counties.

Var. glabrata, Benth. Smooth Wild Mint. Ballast, Camden, C. F. Parker; New Jersey, W. H. Leggett. Rare.

M. rubra, L. Red Mint. Near Phillipsburg, A. P. Garber; in ballast at Camden, C. F. Parker. Adv. Eu.

M. sylvestris, L. Whitehorse, Camden Co., Isaac Burk. Adv. Eu.

Var. alopecuroides, Baker. Hunterdon Co., A. P. Garber; ballast at Camden, I. C. Martindale. Adv. Eu.

Lycopus, L. . . . . WATER HOREHOUND.

L. Virginicus, L. Bugle-weed. Shady moist places. Rather common throughout.

L. sinuatus, Ell. Low grounds. common throughout.

L. sessilifolius, Gray. Swamps near Atsion, W. M. Canby, C. F. Parker; Tom's River, C. F. Parker.

L. Europæus, L. In ballast at Camden, C. F. Parker. Adv. Eu.

Cunila, L. . . . . . . DITTANY.

C. Mariana, L. Common Dittany. Sparingly throughout the State. Weehawken, Torrey Catalogue; Palisades, rare near Closter, C. F. Austin; Snake Hill, W. H. Leggett; First Mt., Essex Co., and Little Falls, Passaic Co., H. H. Rusby; near Keyport, Monmouth Co., R. W. Brown; Mountains at Plainfield, F. Tweedy; near Bridgeton, Cumberland Co., N. L. Britton.

Pycnanthemum, Michx. . . . MOUNTAIN MINT.

P. aristatum, Michx. Sparingly in the pine barrens. Ocean Co., P. D. Knieskern; Monmouth Co., R. W. Brown.

P. incanum, Michx. Not common. Palisades and Closter, Bergen Co., C. F. Austin; Long Hill, W. H. Leggett; Princeton and Paterson, O. R. Willis; First Mt., Essex Co., H. H. Rusby; Water Gap, Warren Co., Camden Co., and Quaker Bridge, Atlantic Co., C. F. Parker; Plainfield, Frank Tweedy.

P. clinopodioides, Torr. & Gray. Rare. Closter, Bergen Co., C. F. Austin.

P. Torreyi, Benth. Closter, Bergen Co., C. F. Austin; Freehold, O. R. Willis.

P. muticum, Pers. Not common. Weehawken, Torrey Catalogue; Closter, C. F. Austin; common in Essex Co., H. H. Rusby; New Dur-

ham and Bergen Point, W. H. Leggett; above Phillipsburg, Warren Co., T. C. Porter.

P. lanceolatum, Pursh. Dry ground. Rather common throughout.

P. linifolium, Pursh. Dry grounds. Common in the middle and southern counties, and sparingly in the northern parts of the State.

Origanum, L. . . . . WILD MARJORAM.

O. vulgare, L. Wild Marjoram. Hoboken, C. F. Austin; Weehawken, W. H. Leggett; Warren Co., F. Knighton. Nat. Eu.

Thymus, L. . . . . THYME.

T. Serpyllum, L. Creeping Thyme. Roadsides, Morris Co., C. F. Austin; Warren Co., F. Knighton. Adv. Eu.

Calamintha, Mœnch. . . . . . . CALAMINTH.

C. Clinopodium, Benth. Basil. Closter, C. F. Austin; First Mt., Essex Co., H. H. Rusby; banks of the Delaware, Warren Co., C. F. Parker; rather rare in Ocean and Monmouth Co., P. D. Knieskern. Probably not native to any part of New Jersey. Nat. Eu.

Melissa, L. . . . . BALM.

M. officinalis, L. Common Balm. Sparingly escaped from gardens. Bergen Point, W. H. Leggett; Weehawken, M. Ruger; Hightstown, O. R. Willis. Nat. Eu.

Hedeoma, Pers. . . . . MOCK PENNYROYAL.

H. pulegioides, Pers. American Pennyroyal. Barren woods and fields. Common throughout the State.

Collinsonia, L. . . . . HORSE-BALM.

C. Canadensis, L. Rich-weed. Stone-root. Rich woods. Common in the northern, and frequent in the middle and southern counties.

Salvia, L. . . . . SAGE.

S. lyrata, L. Lyre-leaved Sage. Frequent on the Yellow Drift and mostly confined to that formation. Sandy fields, New Jersey, Torrey Catalogue; Ocean and Monmouth Cos., P. D. Knieskern; Navesink Highlands, R. W. Brown; Red Bank, Monmouth Co., W. H. Leggett; Camden and Gloucester Cos., C. F. Parker; New Egypt, N. L. Britton.

Monarda, L. . . . . HORSE-MINT.

M. didyma, L. Oswego Tea. Balm. Rare. Bergen Co., C. F. Austin.

M. fistulosa, L.   Wild Bergamot.   Not common.   Woods, New Jersey, Torrey Catalogue; Morris Co., C. F. Austin; near Phillipsburg, T. C. Porter; Franklin, Essex Co., H. H. Rusby; near Keyport, Monmouth Co., R. W. Brown.

M. punctata, L.   Horse Mint.   Abundant on the Yellow Drift and mostly confined to it.

Lophanthus, Benth.   . . . .   GIANT HYSSOP.

L. nepetoides, Benth.   Rare.   Plainfield, F. Tweedy; Weehawken and Hoboken, C. F. Austin; Montclair, Essex Co., W. M. Wolfe; Freehold, O. R. Willis.

L. scrophulariæfolius, Benth.   Not common.   Hoboken, C. F. Austin; Weehawken. N. L. Britton; Freehold, O. R. Willis; banks of the Delaware, near Camden, C. F. Parker.

Physostegia, Benth.   .   . .   FALSE DRAGON-HEAD.

P Virginiana, Benth.   Escaped from cultivation at Plainfield, F. Tweedy.

Nepeta, L. . . . .   CAT-MINT.

N. Cataria, L.   Catnip.   Roadsides and waste places near dwellings. Common throughout.   Adv. Eu.

N. Glechoma, Benth.   Ground Ivy.   Gill.   Waste places.   Quite common throughout.   Adv. Eu.

Brunella, Tourn.   . . . . .   .   SELF-HEAL.

B. vulgaris, L.   Heal-all.   Woods and fields.   Common throughout. Probably introduced from Europe for the most part.   Eu.

Scutellaria, L. .   . . .   SKULLCAP.

S. pilosa, Michx.   Hairy Skullcap.   Dry woods.   Frequent throughout the State.

S. integrifolia, L.   Entire-leaved Skullcap.   Frequent throughout the State.

S. galericulata, L.   Rather common in the northern counties, but rare on the Yellow Drift.

S. lateriflora, L.   Mad-dog Skullcap.   Wet shady places.   Common throughout.

Marrubium, L. . . . .   HOREHOUND.

M. vulgare, L.   Common Horehound.   Frequent in waste places and ballast.   Nat. Eu.

Galeopsis, L. . . . .   HEMP-NETTLE.

G. Tetrahit, L.   Common Hemp-nettle.   In ballast at Camden, C. F. Parker; and Communipaw, A. Brown; Warren Co., F. Knighton. Nat. Eu.

G. Ladanum, L. Red Hemp-nettle. Near dwellings in Ocean and Monmouth Cos., rare, P. D. Knieskern, in Catalogue.

Stachys, L. . . . . HEDGE-NETTLE.
S. arvensis, L. Woundwort. In ballast at Communipaw and Hoboken, A. Brown; and at Camden, C. F. Parker. Adv. Eu.
S. palustris, L. In ballast at Camden, C. F. Parker. Adv. Eu.
S. aspera, Michx. Low grounds. Rather common throughout.
S. hyssopifolia, Michx. Closter, Bergen Co., C. F. Austin; Camden, W. M. Canby; Plainfield, F. Tweedy. Scarce.

Leonurus, L. . . . . MOTHERWORT.
L. Cardiaca, L. Common Motherwort. Waste places. Common throughout. Nat. Eu.

Lamium, L. . . . . DEAD-NETTLE.
L. amplexicaule, L. Frequent in cultivated fields and in ballast. Adv. Eu.
L. purpureum, L. In ballast at Camden, C. F. Parker. Adv. Eu.
L. album, L. In ballast at Camden, C. F. Parker, Adv. Eu.

Ballota, L.   . . . . FETID HOREHOUND.
B. nigra, L. Black Horehound. In ballast at Camden, C. F. Parker. Adv. Eu.

PLANTAGINACEÆ.

Plantago, L. . . . . PLANTAIN. RIBGRASS.
P. major, L. Great Plantain. Waste ground and ballast. Quite common throughout. Nat. Eu.
P. Rugelii, Dec. Common Plantain. Moist or dry soil. Everywhere common. Formerly confounded with the preceding species.
P. maritima, L.; Var. juncoides, Gray. Frequent in salt marshes.
P. lanceolata, L. Ribgrass. Dry fields. Common throughout. Nat. Eu.
P. Virginica, L. Sparingly in the northern counties, but common on the Yellow Drift.
P. pusilla, Nutt. Rare. Cape May, and Ocean Co., C. F. Austin; near Haddonfield, Camden Co., E. Diffenbaugh.
P. heterophylla, Nutt. In ballast at Camden, C. F. Parker.
P. Patagonica, Jacq.; Var. aristata, Gray. "Roadside along the west bank of Maurice River one-half mile below Millville, Cumberland Co.," S. W. Knipe. See Bull. Torr. Bot. Club, Vol. VI., p. 324.

Division C.—Apetalæ.

AMARANTACEÆ.

Amarantus, Tourn. . . . . AMARANTH.

A. hypochondriacus, L. In ballast, and sparingly escaped from gardens. Communipaw, A. Brown; Hunterdon and Warren Cos., F. Knighton. Adv. Tropical America.

A. paniculatus, L. Waste ground, Camden, C. F. Parker. Adv Tropical America.

A. retroflexus, L.; *Var.* chlorostachys, Gray. Waste ground. Common throughout. Adv. Tropical America.

A. albus, L. Waste grounds and roadsides. Common throughout. Nat. Tropical America.

A. spinosus, L. Waste ground and ballast at Camden, C. F. Parker; Ocean and Monmouth Cos., P. D. Knieskern; ballast at Communipaw, A. Brown. Nat. Tropical America.

A. lividus, L. In ballast at Camden, C. F. Parker. Adv. Tropical America.

A. pumilus, Raf. Sparingly in sands of the sea-shore, Sandy Hook to Cape May.

Acnida, L. . . . . WATER HEMP.

A. cannabina, L. Water Hemp. Rather common in salt and brackish marshes along the coast.

CHENOPODIACEÆ.

Chenopodium, L. . . . . . GOOSEFOOT. PIGWEED.

C. polyspermum, L. In ballast at Camden, C. F. Parker; and Communipaw, Addison Brown. Adv. Eu.

C. album, L. Lamb's-quarters. Pigweed. Common in waste and cultivated grounds. Nat. Eu.

C. glaucum, L. Oak-leaved Goosefoot. Hoboken, C. F. Austin; Weehawken, N. L. Britton; Hudson City, M. Ruger; in ballast at Camden, C. F. Parker; and at Communipaw, A. Brown; Phillipsburg, T. C. Porter. Nat. Eu.

C. urbicum, L. Closter, C. F. Austin; Newark, W. H. Leggett; waste grounds and ballast at Camden, C. F. Parker. Nat. Eu.

C. murale, L. In ballast at Camden, C. F. Parker. Adv. Eu.

C. hybridum, L. Maple-leaved Goosefoot. Waste places. Common throughout. Nat. Eu.

10

C. Botrys, L. Jerusalem Oak. Waste places and roadsides. Quite common throughout. Nat. Eu.

C. ambrosioides, L. Mexican Tea. Waste places. Common throughout. Nat. Tropical America.

Var. anthelminticum, Gray. Wormseed. Waste places. Not so common as the type. Nat. Tropical America.

C. multifidum, L. In ballast at Camden, C. F. Parker. Adv. South America.

Blitum, Tourn. . . . . BLITE.

B. maritimum, Nutt. Coast Blite. Sparingly in salt meadows along the coast.

B. capitatum, L. Strawberry Blite. In a garden at Orange, N. J., many years since, W. H. Leggett in Bull. Torr. Bot. Club, II., 44. Eu.

Atriplex, Tourn. . . . . ORACHE.

A. patula, L. Common Orache. Common on salt marshes and brackish river-banks, and very variable. Eu.

A. arenaria, Nutt. Silvery Orache. Frequent on sands of the sea-shore.

A. rosea, L. Red Orache. In ballast at Communipaw and Hoboken, A. Brown; and at Camden, C. F. Parker. Adv. Eu.

Salicornia, Tourn. . . . . SAMPHIRE.

S. herbacea, L. Common on salt marshes Eu.

S. mucronata, Bigel. Common on salt marshes. Eu.

S. ambigua, Michx. (S. fruticosa, L.; *Var.* ambigua, Gray.) Frequent on wet sands of the sea-shore. Eu.

Suæda, Forsk. . . . . SEA BLITE.

S. maritima, Moq. Sea Blite. Rather common on salt marshes. Eu.

Salsola, L. . . . . SALTWORT.

S. Kali, L. Common Saltwort. Sandy sea-shore. Common along the whole coast. Eu.

PHYTOLACCACEÆ.

Phytolacca, Tourn. . . . . POKEWEED.

P. decandra, L. Common Poke or Scoke. Low grounds. Quite common in all parts of the State.

POLYGONACEÆ.

Polygonum, L. . . . . KNOTWEED.

P. orientale, L. Prince's Feather.' Waste places, frequent. Very abundant on river-dredgings at Camden, I. C. Martindale. Adv. India.

P. Careyi, Olney. Carey's Knotweed. Margin of a swamp between Tenafly and Cresskill, 1858, C. F. Austin ; Tom's River, T. C. Porter; Ocean Grove. Monmouth Co., near Winslow, Camden Co., Egg Harbor City and Manchester, Ocean Co., C. F. Parker.

P. Pennsylvanicum, L. Knotweed. Moist open places. Common throughout.

P. incarnatum, Ell. Borders of ponds. Apparently quite scarce. In ballast at Camden, C. F. Parker

P. lapathifolium, Ait., and *Var.* incanum, Gray. In ballast at Camden, C. F. Parker.

P. Persicaria, L. Lady's Thumb. Damp waste places. Common throughout. Nat. Eu.

P. Hydropiper, L. Water-pepper. Wet places. Rather common throughout.

P. acre, H. B. K. Water Smartweed. Wet places. Common throughout.'

P. hydropiperoides, Michx. Mild Water-pepper. Wet places. Common throughout.

P. amphibium, L.; *Var.* aquaticum, Willd. Sparingly in wet places throughout the northern and middle counties.

Var. terrestre, Willd. Closter, Bergen Co., C. F. Austin.

P. Virginianum, L. Woods and thickets. Rather common except in the pine barrens.

P. aviculare, L. Knotgrass. Door-weed. Very common in yards and waste places, and along roadsides in all parts of the State.

P. erectum, L. (P. aviculare, L.; *Var.* erectum, Gray.) Erect Knotgrass. Waste places. Common throughout.

P. maritimum, L. Coast Knotgrass. Frequent in sands of the sea-shore.

P. tenue, Michx. Rocky hills, northern and middle counties. Closter, Bergen Co., C. F. Austin ; Plainfield, Frank Tweedy ; Princeton, O. R. Willis ; Palisades, N. L. Britton ; First Mt., Essex Co., H. H. Rusby ; also at Keyport, Monmouth Co., R. W. Brown.

P. arifolium, L. Halberd-leaved Tear-thumb. Low grounds. Common throughout.

P. sagittatum, L. Arrow-leaved Tear-thumb. Low grounds. Common throughout.

P. Convolvulus, L. Black Bindweed. Waste and cultivated grounds. Quite common. Nat. Eu.

P. dumetorum, L.; *Var*. scandens, Gray. Damp ground. Common throughout.

Polygonella, Mich. (Polygonum, L.)   .   JOINTWEED.

P. articulata, L. Jointweed. Common on the sands of the Yellow Drift, and on the sea-shore.

Fagopyrum, Tourn. . . . . BUCKWHEAT.

F. esculentum, Mœnch. Buckwheat. Frequent in waste and cultivated fields. Adv. Eu.

Rumex, L. . . . . DOCK. SORREL.

R. Patienta, L. Patience Dock. In ballast at Communipaw, A. Brown. Adv. Eu.

R. orbiculatus, Gray. Great Water-dock. Sparingly in the northern counties. Closter, Bergen Co., C. F. Austin; Budd's Lake, T. C. Porter; Secaucus Swamp, W. H. Leggett; Essex Co., H. H. Rusby.

R. Brittanica, L. Pale Dock. Rare. Closter, Bergen Co., C. F. Austin; Hackensack Meadows, W. H. Leggett.

R. verticillatus, L. Swamp Dock. Little Snake Hill, W. H. Leggett.

R. crispus, L. Curled Dock. Waste and cultivated fields. Very common throughout. Nat. Eu.

R. obtusifolius, L. Bitter Dock. Waste and cultivated fields. Quite common throughout. Nat. Eu.

R. sanguineus, L. Bloody-veined Dock. In ballast at Camden, C. F. Parker. Nat. Eu.

R. maritimus, L. Golden Dock. In ballast at Hoboken, Addison Brown. Probably grows in many of the salt marshes along the coast but is not reported from any, except those of Ocean and Monmouth Cos., by Dr. P. D. Knieskern.

R. Acetosella, L. Sheep Sorrel. Old fields and waste places. Common throughout. Nat. Eu.

PODOSTEMACEÆ.

Podostemon, Michx. . . . . RIVER-WEED.

P. ceratophyllus, Michx. River-weed. Sparingly on the bottoms of shallow streams. Passaic River, C. F. Austin; Delaware River above Phillipsburg, T. C. Porter.

ARISTOLOCHIACEÆ.

Asarum, Tourn. . . . . WILD GINGER.

A. Canadense, L.   Wild Ginger.   Camden Co.. C. F. Parker ; Freehold, O. R. Willis; Plainfield, Frank Tweedy; and common in the northern counties.

Aristolochia, Tourn. . . . . BIRTHWORT.

A. Serpentaria, L.   Virginia Snakeroot.   Closter, Bergen Co., C. F. Austin; Weehawken and Long Hill, W. H. Leggett; Hightstown, O. R. Willis; Milburn, Essex Co., N. L. Britton ; Keyport and Holmdel, Monmouth Co., R. W. Brown.

SAURURACEÆ.

Saururus, L. . . . . LIZARD'S-TAIL.

S. cernuus, L.   Nodding Lizard's-tail.   Swamps.   Rather common throughout.

LAURACEÆ.

Sassafras, Nees. . . . .   SASSAFRAS.

S. officinale, Nees.   Common Sassafras.   Woods and copses.   Common throughout.

Lindera, Thunb. . . . . WILD ALLSPICE.

L. Benzoin, Meisn.   Spice-bush.   Low woods.   Quite common throughout.

THYMELEACEÆ.

Dirca, L. . . . .   LEATHERWOOD.

D. palustris, L.   Moosewood.   Rare.   Shady woods, New Jersey, Torrey Catalogue ; Closter and Palisades, Bergen Co., C. F. Austin.

LORANTHACEÆ.

Phoradendron, Nutt. .   . . .   FALSE MISTLETOE.

P. flavescens, Nutt.   American Mistletoe.   Sparingly on the Yellow Drift formation, generally parasitic on the Black Gum, but occasionally on other deciduous-leaved trees.   Never on Conifers.   Between Manchester and Lakewood, Ocean Co., Wm. Bower ; Kaighn's Swamp, Camden, I. C. Martindale ; Jackson, Camden Co., C. F. Parker, (on the Red Maple); Hightstown, O. R. Willis ; Medford and New Lisbon, Burlington Co., N. L. Britton.

SANTALACEÆ.

Comandra, Nutt. . . . . BASTARD TOAD-FLAX.

C. umbellata, Nutt. Dry grounds. Rather common in all parts
of the State. According to the observations of Mr. Jos. Schrenck this
plant is generally parasitic on species of Vaccinium or Gaylussacia.
See Bulletin Torrey Botanical Club, vol. vii., p. 67, 1880.

EUPHORBIACEÆ.

Euphorbia, L. . . . . SPURGE.

E. polygonifolia, L.   Common in sands of the sea-shore.
E. serpens, H. B. K. In ballast at Camden, C. F. Parker.
E. maculata, L.   Fields and roadsides. Very common throughout.
E. hypericifolia, L.   Field and roadsides. Quite common through-
out.
E. corollata, L.   Quite common on the Yellow Drift, and mostly
confined to that formation.
E. Ipecacuanhæ, L.   Common on the Yellow Drift, and mostly con-
fined to that formation.
E. Helioscopia, L. In ballast at Camden, C. F. Parker. Nat. Eu.
E. Peplus, L.   Warren Co., F. Knighton; in ballast at Camden, C.
F. Parker. Adv. Eu.
E. Lathyris, L. Caper Spurge. Mohinkson Hill, near Keyport, S.
Lockwood. Adv. Eu.
E. Cyparissias, L. Frequently escaped from cultivation. Essex Co.,
H. H. Rusby; Camden Co., D. G. Brinton; Closter, Bergen Co., C. F.,
Austin; Chatham, W. H. Leggett; borders of Greenwood Lake, Red
Bank, Monmouth Co., W. H. Rudkin: Newton, Sussex Co., L. Schöney;
Plainfield, F. Tweedy. Adv. Eu.

Acalypha, L. .   . THREE-SEEDED MERCURY.

A. Virginica, L. Fields and open places. Common throughout.
Var. gracilens, Gray. Common on the Yellow Drift, and sparingly
in other districts.
A. Caroliniana, Walt. Rare. Princeton, Dr. John Torrey; near
Trenton, Isaac Burk; Closter, C. F. Austin

Croton, L. . . . . CROTON.

C. glandulosus, L. In ballast at Camden, C. F. Parker.
C. capitatus, Michx. Pine barrens of New Jersey, P. D. Knieskern
in Gray's Manual, p. 438.

Crotonopsis, Michx. . . . . 'CROTONOPSIS.

C. linearis, Michx. Sparingly on the sands of the Yellow Drift. Near Manchester, Ocean Co., and Southwark, P. D. Knieskern; Gloucester Co., C. F. Parker.

URTICACEÆ.

Ulmus, L. . . . . ELM.

U. fulva, Michx. Slippery Elm. Frequent in the middle and northern counties. Closter, Bergen Co., C. F. Austin; Plainfield, Frank Tweedy; Palisades, W. H. Leggett; Newton, Sussex Co., N. L. Britton; First Mt., Essex Co., H. H. Rusby; Mercer and Monmouth Cos., O. R. Willis.

U. Americana, L. American Elm. River banks and low grounds. Common except in the pine barrens.

Celtis, Tourn. . . . . HACKBERRY.

C. occidentalis, L. Hackberry. Frequent throughout the State. Banks of the Delaware, Camden, C. F. Parker; banks of Tom's River, Ocean Co., P. D. Knieskern; Essex Co., H. H. Rusby; Closter, scarce, C. F. Austin; Keyport and Union, Monmouth Co., S. Lockwood; Weehawken, W. H. Leggett; near Newton, Sussex Co., Arthur Hollick; Fort Lee, N. L. Britton; Plainfield, F. Tweedy.

Morus, Tourn. . . . . MULBERRY.

M. rubra, L. Red Mulberry. Closter, Bergen Co., C. F. Austin; Chatham, W. H. Leggett; Hoboken Heights, M. Ruger; Plainfield, Frank Tweedy; Keyport, Monmouth Co., R. W. Brown. Not common.

M. alba, L. White Mulberry. Sparingly escaped from cultivation. Camden, C. F. Parker; Monmouth Co., R. W. Brown; Hoboken, W. H. Leggett; Bridgeton, N. L. Britton. Adv. Eu.

Urtica, Tourn. . . . . NETTLE.

U. gracilis, Ait. Sparingly in the northern counties. Hackensack, C. F. Austin; Warren Co., F. Knighton; Essex Co., H. H. Rusby.

U. dioica, L. Waste grounds and banks of the Delaware at Camden, C. F. Parker; Warren Co., F. Knighton; Bergen Hill, M. Ruger; Essex Co., H. H. Rusby. Nat. Eu.

U. urens, L. In ballast at Communipaw, A. Brown; and Camden, C. F. Parker. Adv. Eu.

Laportea, Gaud. . . . . . WOOD-NETTLE.

L. Canadensis, Gaud. Damp woods. Quite common in the northern and middle counties.

Pilea, Lindl. . . . . . RICHWEED. CLEARWEED.

P. pumila, Gray. Richweed. Damp woods. Common in the northern and middle counties.

Bœhmeria, Jacq. . . . . FALSE NETTLE.

B. cylindrica, Willd. Damp woods. Common throughout.

Parietaria, Tourn. . . . . PELLITORY.

P. Pennsylvanica, Muhl. Rare. Mercer Co., Dr. John Torrey; Closter, Bergen Co., C. F. Austin; Sandy Hook, 1870, M. Ruger.

Cannabis, Tourn. . . . . HEMP.

C. sativa, L. Common Hemp. Waste ground, Camden, C. F. Parker; Keyport, Monmouth Co., R. W. Brown; Communipaw, W. H. Leggett; Essex Co., H. H. Rusby. Adv. Eu.

Humulus, L. . . . . HOP.

H. Lupulus, L. Common Hop. Chatham, W. H. Leggett; Hoboken Heights, M. Ruger; near Washington, Warren Co., C. F. Parker; Closter, Bergen Co., C. F. Austin. Also escaped from cultivation in many other places.

PLATANACEÆ.

Platanus, L. . . . . PLANE-TREE.

P. occidentalis, L. Buttonwood. Alluvial banks. Quite common throughout the State.

JUGLANDACEÆ.

Juglans, L. . . . . WALNUT.

J. cinerea, L. Butternut. Frequent in the northern and middle counties. Also Ocean and Monmouth Cos., P. D. Knieskern. Rare in the southern parts of the State.

J. nigra, L. Black Walnut. Low woods. Frequent except in the pine barren districts.

Carya, Nutt. . . . . HICKORY.

C. alba, Nutt. Shag-bark Hickory. Woods. Not rare, except in the pine barrens.

C. microcarpa, Nutt. Small-fruited Hickory. Rare. Closter, Bergen Co., C. F. Austin.

C. tomentosa, Nutt. Mocker-nut. Woods. Common throughout.

C. porcina, Nutt. Pig-nut. Rather common in most parts of the State.

C. amara, Nutt. Low woods. Rather common.

MYRICACEÆ.

Myrica, L . . . . . . BAYBERRY. WAX-MYRTLE.

M. Gale, L. Sweet Gale. Warren Co., F. Knighton, in Willis Catalogue.

M. cerifera, L. Bayberry. Light sandy soil. Common throughout the State.

Comptonia, Solander. . . . . SWEET-FERN.

C. asplenifolia, Ait. Sweet Fern. Dry sandy soil. Common throughout the State.

CUPULIFERÆ.

Quercus, L. . . . . OAK.

Q. alba, L. White Oak. Woods. Common throughout.

Q. obtusiloba, Michx. Post Oak. Bergen Co., C. F. Austin ; common on the Yellow Drift.

Q. macrocarpa, Michx. Mossy-cup Oak. Rare. Quaker Bridge, Atlantic Co , I. C. Martindale.

Q. bicolor, Willd. Swamp White Oak. Low grounds. Frequent in the northern counties.

Q. Prinus, L. Chestnut Oak. Rather common in all parts of the State, but whether it is the typical form, or one or both varieties is yet to be determined. I have what seems to be *Var.* acuminata, Michx., from the Palisades, and Mr. Addison Brown has *Var.* monticola, Michx., from Neversink Highlands. One or other of the varieties is abundant on Little Snake Hill. Botanists will oblige me by making this point a subject of special study during the coming season, and I shall be grateful for specimens from all parts of the State. The acorns are indispensable to a satisfactory determination of the varieties.

Q. prinoides, Willd. Chinquapin Oak. Sparingly on the Yellow Drift. Ocean Co., C. F. Parker; Bridgeton, Cumberland Co., N. L. Britton.

Q. Phellos, L. Willow Oak. Frequent on the Yellow Drift and confined to that formation. Keyport, S. Lockwood, N. L. Britton ; Camden and Gloucester Cos., C. F. Parker ; near Long Branch, W. H.

Leggett; South River, Middlesex Co., N. L. Britton. Not rare in the southern counties.

Q. imbricaria, Michx. Shingle Oak. Sparingly in pine barren regions. Definite localities are desired.

Q. nigra, L. Black Jack Oak. Common on the Yellow Drift and mostly confined to that formation.

Var. quinqueloba, Eng. Tom's River, Ocean Co., !. C. Martindale.

Q. heterophylla, Michx. Bartram's Oak. Near Woodbury, Gloucester Co., C. F. Parker; Cape May Co., C. F. Austin.

Q. ilicifolia, Wang. Black Scrub Oak. Common, especially in the pine barrens.

Q. falcata, Michx. Spanish Oak. Not common, and confined to the Yellow Drift formation. Cape May, C. F. Austin; Keyport, Monmouth Co., S. Lockwood; Point Pleasant, Ocean Co., P. D. Knieskern; Camden and Gloucester Cos., C. F. Parker; Bridgeton, Cumberland Co., N. L. Britton.

Q. coccinea, Wang. Scarlet Oak. Frequent throughout the State.

Var. tinctoria, Gray. Black Oak. Woods. Common throughout.

Q. rubra, L. Red Oak. Woods. Common throughout.

Q. palustris, Du Roi. Pin Oak. Swampy and low ground. Rather common throughout.

Castanea, Tourn. . . . . CHESTNUT.

C. vesca, L.; *Var.* Americana, Michx. Chestnut. Woods. Common throughout the State.

C. pumila, Michx. Chinquapin. Growing abundantly at Clarksboro, Gloucester Co., Isaac Burk.

Fagus. Tourn. . . . . BEECH.

F. ferruginea, Ait. American Beech. Common in the northern and middle counties, and frequent on the Yellow Drift.

Corylus, Tourn. . . . . HAZEL-NUT.

C. Americana, Walt. Wild Hazel-nut. Woods and thickets. Common throughout.

C. rostrata, Ait. Beaked Hazle-nut. Sparingly in the northern and middle counties. Mercer Co., Dr. John Torrey; Warren Co., C. F. Parker; Morris Co., C. F. Austin; Phillipsburg, T. C. Porter; Plainfield, Frank Tweedy; New Providence, W. H. Leggett; Bloomsbury, Hunterdon Co., A. P. Garber.

Ostrya, Michx. . . . . HOP-HORNBEAM.

O. Virginica, Willd. American Hop-hornbeam. Sparingly in the northern counties. Closter, Bergen Co., C. F. Austin; Palisades, W. H. Leggett.

Carpinus, L. . .   . .   IRON-WOOD.
C. Americana, Michx.   American Hornbeam.   River banks.
Rather common throughout the State.

Betula, Tourn. . . . .   BIRCH.
B. lenta, L.   Sweet or Black Birch.   Common in the northern and
middle counties.
B. lutea, Michx., f.   Yellow Birch.   Sparingly in the northern coun-
ties. Closter. Bergen Co., C. F. Austin ; Essex Co., H. H. Rusby.
B. alba, L.; Var. populifolia, Spach.   White Birch.   Low grounds.
Common throughout.
B. nigra, L.   River or Red Birch.   Closter, Bergen Co., C. F. Austin;
banks of the Delaware, Camden, C. F. Parker; Bull's Ferry, P. V.
Le Roy; Paterson, W. H. Leggett; Plainfield, F. Tweedy.
B. pumila, L.   Low Beech.   Sparingly in the northern counties.
Newton, Sussex Co., A. P. Garber; Budd's Lake, Morris Co., C. F.
Parker, T. C. Porter.

Alnus, Tourn. . .   . . ALDER.
A. incana, Willd.   Hoary Alder.   Budd's Lake, Morris Co., C. F.
Austin. Eu.
A. serrulata, Ait.   Smooth Alder.   Low grounds.   Common through-
out.

SALICACEÆ.

Salix, Tourn.   . . .   WILLOW. OSIER.
S. candida, Willd.   Hoary Willow. . Rare.   Budd's Lake, Morris
Co., T. C. Porter.   .
S. tristis, Ait.   Dwarf Gray Willow.   Frequent.   Bergen Co., C. F.
Austin; Gloucester Co., C. E. Smith ; Camden Co., C. F. Parker ; South
Amboy, W. H. Leggett.
S. humilis, Marsh.   Prairie Willow.   Dry fields. Quite common
throughout.
S. discolor, Muhl.   Glaucous Willow.   Low grounds.   Common
throughout.
S. sericea, Marsh.   Silky Willow.   Low grounds.   Not rare.
S. cordata, Muhl.   Heart leaved Willow   Low grounds.   Frequent.
S. viminalis, L.   Basket Osier.   Cultivated in New Jersey, Bull.
Torr. Bot. Club, Vol. III., p. 44 ; Warren Co., F. Knighton.   Adv. Eu.
S. livida, Wahl.; Var. occidentalis, Gray.   Livid Willow.   Spar-
ingly in the northern counties. Palisades, C. F. Austin ; Warren Co.,
A. P. Garber, T. C. Porter.

.

S. lucida, Muhl. Shining Willow. Andover, Sussex Co., A. P. Garber, C. F. Austin; Budd's Lake, Morris Co., T. C. Porter; Verona, Essex Co., H. H. Rusby.

S. nigra, Marsh. Black Willow. Frequent in the southern and middle counties.

S. fragilis, L. Brittle Willow. Little Falls, Passaic Co., H. H. Rusby; Delaware Water Gap, A. P. Garber; Camden, C. F. Parker. Adv. Eu.

S. alba, L. White Willow. Common along streams in all parts of the State. Adv. Eu.

S. longifolia, Muhl. Long-leaved Willow. Delaware Water Gap, A. P. Garber; Marble Hill, Warren Co., T. C. Porter; banks of the Delaware at Camden, C. F. Parker.

S. myrtilloides, L. Myrtle Willow. Budd's Lake, Morris Co., T. C. Porter, C. F. Austin. Eu.

Populus, Tourn. . . . . .   . POPLAR. ASPEN.

P. tremuloides, Michx. American Aspen. Woods. Rather common throughout.

P. grandidentata, Michx. Large-toothed Aspen. Frequent in the northern and middle counties.

P. heterophylla, Ait. Downy Poplar. Rare. "Found in N. J. June 6th, 1814," in Herb. Acad. Nat. Sci. Phil.

P. angulata, Ait. Angled Cottonwood. Above Phillipsburg, Warren Co., T. C. Porter; Holland Station, Hunterdon Co., C. F. Parker.

EMPETRACEÆ.

Corema, Don. . . . . BROOM CROWBERRY.

C. Conradii, Torr. In pine barren regions, but very scarce. Cedar Creek, Ocean Co., Dr. John Torrey, but was not found there by Messrs. J. H. Redfield and C. F. Parker, who went in search of it; a reported locality at Pemberton, Burlington Co., proved equally disappointing; borders of pine woods, Ocean and Monmouth Cos., O. R. Willis in Catalogue; New Jersey, Gray's Manual.

CERATOPHYLLACEÆ.

Ceratophyllum, L. . . . . HORNWORT.

C. demersum, L. Hornwort. Frequent in slow streams. Bergen Co., C. F. Austin; tidal mud, Delaware River at Camden, C. F. Parker. Eu.

Sub-Class 2.—GYMNOSPERMÆ.

CONIFERÆ.

Pinus, Tourn. . . . . Pine.

P. rigida, Miller. Pitch Pine. Sandy or rocky soil. Common in all parts of the State, but particularly abundant on the Yellow Drift where it forms the forests of the pine barrens.

P. inops, Ait. Scrub Pine. Frequent in the southern and middle counties, but scarce in pine barren regions. Camden Co., C. F. Parker; Monmouth, Ocean and Burlington Cos., O. R. Willis; Milford, Hunterdon Co., and Bridgeton, Cumberland Co., N. L. Britton.

P. mitis, Michx. Yellow Pine. Scarce. Pine barrens, Atlantic Co., and Burlington Co., I. C. Martindale; Camden Co., C. F. Parker; Ocean and Monmouth Cos., P. D. Knieskern.

P. Strobus, L. White Pine. Frequent in the northern and middle counties.

Abies, Tourn. . . . . Spruce. Fir.

A. nigra, Poir. Black Spruce. Sparingly in the northern counties. Budd's Lake, Morris Co., C. F. Parker; Hunterdon Co., F. Knighton; New Durham Swamp, Torrey Catalogue; Secaucus Swamp, W. H. Leggett; Palisades, rare, C. F. Austin.

A. Canadensis, Michx. Hemlock Spruce. Common in woods in the northern counties. Scarce elsewhere.

Larix, Tourn. . . . . Larch.

L. Americana, Michx. American Larch. Tamarack. Sparingly in the northern counties. Budd's Lake, Morris Co., T. C. Porter; New Durham Swamp, W. H. Leggett; Passaic River and Closter, Bergen Co., C. F. Austin.

Thuja, Tourn. . . . . Arbor Vitæ.

T. occidentalis, L. American Arbor Vitæ. Sparingly in the northern counties. Rocky banks of the Hudson, New Jersey, Torrey Catalogue; Closter, Bergen Co., C. F. Austin; Warren Co., F. Knighton.

Cupressus, Tourn. . . . . Cypress.

C. thyoides, L. White Cedar. Very common in swamps on the Yellow Drift, and sparingly in other districts. New Durham Swamp, C. F. Austin; Secaucus Swamp, W. H. Leggett.

Juniperus, L. . . . . CEDAR. JUNIPER.

J. communis, L. Common Juniper. Frequent in the northern and middle counties; rare or absent on the Yellow Drift. Eu.

J. Virginiana, L. Red Cedar. Dry sterile soil. Common in all parts of the State, except in the pine barrens.

Taxus, Tourn. . . . . YEW.

T. baccata, L.; *Var.* Canadensis, Gray. American Yew. Sparingly in the northern counties. Palisades, common, C. F. Austin; Greenwood Lake, W. H. Rudkin; eastern shore of Swartswood Lake, Wm. Bower; on an island in Swartswood Lake. T. C. Porter.

CLASS II.—ENDOGENOUS PLANTS.

HYDROCHARIDEÆ

Anacharis, Richard. . . . . . . WATER-WEED.

A. Canadensis, Planchon. Water-weed. Ponds and slow streams. Rather common throughout.

Vallisneria, Micheli. . . . . . . TAPE-GRASS.

V. spiralis, L. Eel-grass. Streams. Frequent in the northern and middle counties, and sparingly in the southern parts of the State. Eu.

Limnobium, Richard. . . . AMERICAN FROG'S-BIT.

L. Spongia, Richard. Swimming River. Monmouth Co., rare, P. D. Knieskern in Cat. Plants Monmouth and Ocean Cos.

ORCHIDEÆ.

Orchis, L. . . . . ORCHIS.

O. spectabilis, L. Showy Orchis. Sparingly in the northern and middle counties. Hills back of Hoboken, Torrey Catalogue; New Durham, W. H. Leggett; Montclair, Essex Co., H. H. Rusby; Marble Hill, Warren Co., C. F. Parker; Keyport, R. W. Brown; near Freehold, O. R. Willis; Little Falls, Passaic Co., W. M. Wolfe.

Habenaria, Willd., R. Br. . . . . REIN ORCHIS.

H. tridentata. Hook. Sparingly throughout the State. Closter, Bergen Co., C. F. Austin; New Durham Swamp, Torrey Catalogue;

Secaucus Swamp, and Great Swamp near Chatham, W. H. Leggett;
Monmouth and Ocean Cos., P. D. Knieskern; Seabright, Monmouth Co.,
A. Brown; Quaker Bridge, Atlantic Co., Cape May and Camden Cos.,
C. F. Parker. Frequent in pine barren regions.

H. integra, Spreng. Sparingly in pine barren regions. Ocean and
Monmouth Cos., P. D. Knieskern; Quaker Bridge, Atlantic Co., C. F.
Parker; Allaire, S. Lockwood.

H. virescens, Spreng. Frequent in the northern and middle counties. Closter, Bergen Co., C. F. Austin; Secaucus and Chatham, W.
H. Leggett; Millburn, Essex Co., N. L. Britton.

H. viridis, R. Br.; *Var.* bracteata, Reichenbach. Closter, Bergen
Co., rare, C. F. Austin ; Cooper's Furnace, A. P. Garber. Rare, and
confined to the northern counties. Eu.

H. Hookeri, Torr. Sussex Co., C. F. Austin, in Willis Catalogue.

H. orbiculata, Torr. Closter, Bergen Co., and near Sparta, Sussex
Co., C. F. Austin; Newton, Sussex Co., A. P. Garber. Rare, and confined to the northern counties.

H. cristata, R. Br. New Durham Swamp, Torrey Catalogue; Secaucus, Wm. Bower; and in the southern counties. Egg Harbor City
and Cumberland Co., C. F. Parker; Atsion, W. M. Canby.

H. ciliaris, R. Br. Sparingly throughout the State. Hoboken
Meadows, Torrey Catalogue; Tenafly and Closter, C. F. Austin; Red
Bank, Monmouth Co., W. H. Leggett; Keyport, R. W. Brown ; Mon-
mouth and Ocean Cos., P. D. Knieskern ; Atlantic and Camden Cos.,
C. F. Parker; Newfield, Gloucester Co., J. B. Ellis; Franklin, Essex
Co., rare, H. H. Rusby.

H. blephariglottis, Hook. Closter, Bergen Co., C. F. Austin ; Secaucus, W. H. Leggett; and frequent in swamps on the Yellow Drift.

H. lacera, R. Br. Rather frequent in swamps throughout the State.

H. psycodes, Gray. Frequent in the northern and middle counties,
but rare on the Yellow Drift. Camden, C. F. Parker; Freehold, S.
Lockwood.

H. fimbriata, R. Br. Meadows about Elizabethtown, Eddy in Torrey
Catalogue; in a bog at the foot of east side of Copperas Mt., Morris
Co., W. H. Rudkin ; Budd's Lake, Morris Co., C. F. Parker.

H. peramœna, Gray. A single specimen from near Haddonfield,
Camden Co., I. C. Martindale in Bull. Torr. Bot. Club, vi., 331, 1879 ;
near Lawrenceville, Monmouth Co., Lanning in Willis Catalogue.

    Goodyera, R. Br. . . . .   RATTLESNAKE-PLANTAIN.

G. pubescens, R. Br. Rich woods. Common throughout, except
in the pine barrens.

Spiranthes. Rich. . . . . LADIES' TRESSES.

S. latifolia, Torr. Rare. Newton, Sussex Co., A. P. Garber; banks of the Delaware above Burlington, Isaac Burk.

S. cernua, Richard. Wet places. Common throughout.

S. graminea, Lindl.; *Var.* Walteri, Gray. Sparingly throughout the State. Summit, W. H. Leggett; Closter, Bergen Co., and Morris Co., C. F. Austin; Quaker Bridge, Atlantic Co., and Cape May, C. F. Parker; Essex Co., H. H. Rusby.

S. gracilis, Bigelow. Sandy woods and fields. Common throughout the State.

S. simplex, Gray. Rare. Closter, Bergen Co., C. F. Austin; common about Keyport, R. W. Brown; Camden, C. F. Parker.

Listera, R. Br. . . . . TWAYBLADE.

L. cordata, R. Br. Mercer Co., Dr. John Torrey in Willis Catalogue. Eu.

L. australis, Lindl. Damp thickets near Camden; Herb. Acad. Nat. Sci. Phil.; also W. M. Canby in Willis Catalogue; New Brunswick, S. Lockwood.

L. convallarioides, Hook. In the cedar swamp at New Durham, Torrey Catalogue, 1819. Not since found.

Arethusa, Gronov. . . . . ARETHUSA.

A. bulbosa, L. Sparingly throughout the State. Closter, Bergen Co., C. F. Austin; New Durham Swamp, formerly, W. H. Leggett; cedar swamp at Weehawken, Torrey Catalogue; Montclair, Essex Co., Randall Spaulding; frequent about Keyport, R. W. Brown; Freehold, O. R. Willis; Burlington Co., C. F. Parker; Ocean and Monmouth Cos., P. D. Knieskern; Budd's Lake. Morris Co., T. C. Porter; Andover, Sussex Co., A. P. Garber.

Pogonia, Juss. . . . . POGONIA.

P. ophioglossoides, Nutt. Bogs. Sparingly in the northern counties. Closter, C. F. Austin; Milburn, Essex Co., H. H. Rusby; Budd's Lake, Morris Co., C. F. Parker; common in the southern and middle counties.

P. pendula, Lindl. Rare. Closter, and on Palisades, Bergen Co., C. F. Austin; Fort Lee, Wm. Bower.

P. divaricata, R. Br. Very sparingly in the pine barrens. Quaker Bridge, W. H. Leggett; Batsto, Atlantic Co., D. C. Eaton.

P. verticillata, Nutt. Sparingly throughout the State. Ocean and Monmouth Cos., not common, P. D. Knieskern; Hightstown, S. Lockwood; Keyport, R. W. Brown; near Freehold, O. R. Willis; Camden,

C. F. Parker; Franklin, Essex Co., H. H. Rusby; Tenafly, G. I. Cook; rare at Closter, Bergen Co., C. F. Austin.

P. affinis, C. F. Austin. Formerly (1858) grew near Closter and Norwood, Bergen Co,; now (1873) exterminated, C. F. Austin.

Calopogon, R. Br. . . . . CALOPOGON.

C. pulchellus, R. Br. Plant with the same range as Pogonia ophioglossoides, Nutt., and generally growing with it.

Tipularia, Nutt. . . . . . CRANE-FLY ORCHIS.

T. discolor, Nutt. Scarce. Near Newark, Wm. Bower; Bergen Point and Chatham, W. H. Leggett; Great Swamp near Madison, F. J. Bumstead; formerly grew near Closter, C. F. Austin; near Freehold, O. R. Willis.

Microstylis, Nutt. . . ADDER'S-MOUTH.

M. ophioglossoides, Nutt. Rare. Closter, C. F. Austin; New Durham Swamp, Torrey Catalogue; Mercer Co., Dr. John Torrey; Red Bank, Monmouth Co., W. H. Leggett; Keyport, R. W. Brown.

Liparis, Richard. . . . . TWAYBLADE.

L. liliifolia, Richard. Scarce. Weehawken, Torrey Catalogue; Closter and New Durham, C. F. Austin; Marble Hill, Warren Co. and Camden Co., C. F. Parker; Manchester, Ocean Co., P. D. Knieskern; Keyport, R. W. Brown; First Mt., Essex Co., H. H. Rusby; and Plainfield, Frank Tweedy.

L. Lœselii, Richard. Rare, and confined to the northern parts of the State. In a bog on Weehawken Heights, I. H. Hall; Marble Hill, Warren Co., T. C. Porter; Closter, rare, C. F. Austin; First Mt., Essex Co., H. H. Rusby. Eu.

Corallorhiza, Haller. . . . . CORAL-ROOT.

C. innata, R. Br. Rare, and confined to the northern counties. Closter, Bergen Co., C. F. Austin; Blairstown, Warren Co., H. H. Rusby. Eu.

C. odontorhiza, Nutt. Not uncommon in the northern and middle counties. Closter, C. F. Austin, Chatham, W. H. Leggett; near Princeton, O. R. Willis; Newton, Sussex Co., A. P. Garber; Verona, Essex Co., H. H. Rusby; Camden Co., C. F. Parker.

C. multiflora, Nutt. Not uncommon in the northern and middle counties. Closter, C. F. Austin; Newton, Sussex Co., A. P. Garber; near Princeton, O. R. Willis; Blairstown, Warren Co., H. H. Rusby.

Aplectrum, Nutt. . .   Putty-root.   Adam-and-Eve.

A. hyemale, Nutt.   Scarce.   New Jersey, Torrey Catalogue; Closter, C. F. Austin; Palisades, C. F. Austin, C. F. Parker; Newton, Sussex Co., A. P. Garber; Great Notch, Passaic Co., W. M. Wolfe.

Cypripedium, L. . . . .   Lady's Slipper.

C. parviflorum, Salisb.   Smaller Yellow L.   Scarce.   Palisades, rare, C. F. Austin; Tenafly, G. I. Cook, Arthur Hollick; Keyport, Monmouth Co., R. W. Brown; Newton, Sussex Co., A. P. Garber; Warren Co., F. Knighton; First and Second Mts., Essex Co., and High Mt., northwest of Paterson, H. H. Rusby.

C. pubescens Willd.   Larger Yellow L.   New Jersey, Torrey Catalogue; quite common on mountains at Montclair, Essex Co., H. H. Rusby; Carpentersville, Warren Co., A. P. Garber; Little Falls, Passaic Co., W. M. Wolfe; near Englishtown, Middlesex Co., O. R. Willis in Catalogue.

C. spectabile, Swartz.   Showy L.   Very scarce   In the cedar swamp near Weehawken, Torrey Catalogue; Allamuchy Swamp, T. C. Porter; Sparta, Sussex Co., C. F. Austin.

C. acaule, Ait.   Stemless Lady's-slipper.   Sandy woods.   Frequent in all parts of the State.

IRIDEÆ.

Iris, L. . .   . .   Flower-de-Luce.

I. versicolor, L.   Larger Blue Flag.   Wet places.   Common throughout.

I. Virginica, L.   Slender Blue Flag.   Frequent.   Closter, common, C. F. Austin; Hackensack Meadows, W. H. Leggett; New Durham, among rocks, M. Ruger; Plainfield, F. Tweedy; Roseland, Essex Co., H. H. Rusby; Camden Co., C. F. Parker; and common in swamps in the pine barrens.

Pardanthus, Ker. . . . . .   Blackberry-lily.

P. Chinensis, Ker.   Sparingly escaping from gardens.   Preakness, Passaic Co., W. L. Fischer; and reported from near Princeton.   Adv. China.

Sisyrinchium, L. . . . .   Blue-eyed Grass.

S. Bermudiana, L.   Grassy meadows.   Common throughout the State.

HÆMODORACEÆ.

Lacnanthes, L. . .   . . .   Red-root.

L. tinctoria, Ell.   Frequent in swamps in the pine barrens, and confined to the Yellow Drift.

Lophiola, Ker. . . . . LOPHIOLA.

L. aurea, Ker. Common in swamps in the pine barrens, and confined to the Yellow Drift.

Aletris, L. . . . . COLIC-ROOT. STAR-GRASS.

A. farinosa, L. Colic-root. Sandy woods and fields. Sparingly in the northern, but common in the southern and middle counties. Closter, Bergen Co., C. F. Austin; Little Ferry, W. H. Rudkin; Chatham, W. H. Leggett.

A. aurea, Walt. Star-grass. Pine barren regions, but scarce, and definite localities are desired.

HYPOXIDÆ.

Hypoxys, L. . . . . STAR-GRASS.

H. erecta, L. Star-grass. Meadows and open woods. Common in all parts of the State.

DIOSCOREÆ.

Dioscorea, Plum. . . . . YAM-ROOT.

D. villosa, L. Wild Yam-root. Damp thickets. Common throughout.

ALISMACEÆ.

Alisma, L. . . . . WATER PLANTAIN.

A. Plantago, L.; *Var.* Americanum, Gray. Borders of ponds and shallow water. Common throughout.

Sagittaria, L. . . . . ARROW-HEAD.

S. variabilis, Engelm. Ponds, &c. Common throughout, and very variable.

S. calycina, Engelm. Hackensack Marshes and Fairview, Bergen Co., C. F. Austin.

Var. spongiosa, Gray. In tidal mud at Camden, C. F. Parker.

S. heterophylla, Pursh. In wet places at Camden, frequent; W. M. Canby in Willis Catalogue.

S. pusilla, Nutt. Hackensack streams, Torrey Catalogue; Fairview, Bergen Co., C. F. Austin; Passaic, W. H. Leggett; shores of the Delaware River at Camden, C. F. Parker.

JUNCAGINEÆ.

Scheuchzeria, L. . . . . SCHEUCHZERIA.

S. palustris, L. Bogs. Budd's Lake, Morris Co., T. C. Porter; Longacoming, Camden Co., C. E. Smith. Eu.

POTAMEÆ.

Potamogeton, Tourn. . . . . . PONDWEED.

P. natans, L. Ponds. Quite common throughout. Eu.

P. Claytonii, Tuckerman. Ponds and slow streams. Frequent.

P. Spirallus, Tuckerman. (?) Tidal mud, Delaware River. Camden, C. F. Parker.

P. hybridus, Michx. Ponds and streams. Frequent.

P. pulcher, Tuckerman. Scarce. Shallow pond, Atlantic City and near Elmer, Salem Co., C. F. Parker; Tom's River, T. C. Porter.

P. amplifolius, Tuckerman. Andover and Swartswood Lake, T. C. Porter; Hackensack River, C. F. Austin; near Waterloo, Sussex Co., A. P. Garber; tidal mud, Delaware River at Camden, C. F. Parker; upper Passaic River, H. H. Rusby.

P. gramineus, L. Waters of the Hackensack River, C. F. Austin in Bull. Torr. Bot. Club, III., 53. Eu.

P. lucens, L. Waters of the Hackensack River, C. F. Austin in Bull. Torr. Bot. Club, III., 53; upper part of Monmouth Co., O. R. Willis in Catalogue. Eu.

P. perfoliatus, L. Ponds and slow streams. Quite common throughout. Eu.

P. crispus, L. Morris Co., C. F. Austin; Passaic, W. H. Leggett; Lake Hopatcong and Musconetcong Creek, T. C. Porter; streams at Plainfield, common, Frank Tweedy; ditches and tidal mud, Delaware River, Camden, C. F. Parker. Eu.

Var. gemmiparus, Morong. Ditches. Camden, I. C. Martindale.

P. compressus, L. Rare. Hackensack River, C. F. Austin; near Waterloo Station, M. & E. R. R., and Musconetcong Creek, T. C. Porter. Eu.

P. pauciflorus, Pursh. Ponds or slow streams. Frequent.

P. pusillus, L. Hackensack River and tributary creeks, W. H. Leggett; Budd's Lake, Morris Co., T. C. Porter. Probably quite common.

P. Tuckermani, Robbins. Ponds, pine barrens, C. F. Austin; near Atsion, W. M. Canby. Rare.

P. pectinatus, L. Hackensack River, C. F. Austin; Monmouth Co., Dr. John Torrey in Willis Catalogue. Eu.

P. Robbinsii, Oakes. Scarce. Hackensack River, C. F. Austin; Budd's Lake, and Swartswood Lake, Morris Co., T. C. Porter.

Ruppia, L. . . . . DITCH-GRASS.

R. maritima, L. Shallow brackish water along the coast, but scarce. Tom's River, P. D. Knieskern. Eu.

Zannichellia, Michx. . . . . . HORNED POND-WEED.
Z. palustris, L.; *Var.* pedunculata, Gray. Ditches, Bergen Co., C. F. Austin. Very rare. Eu.

NAIADEÆ.

Naias. L. . . . . NAIAD.
N. flexilis, Rostk. Ponds and slow streams. Probably common throughout. Eu.

Zostera, L. . . . . . . GRASS-WRACK. EEL-GRASS.
Z. marina, L. Eel-grass. Common in salt water along the coast. Eu.

TYPHACEÆ.

Typha, Tourn. . . . . CAT-TAIL FLAG.
T. latifolia, L. Common Cat-tail. Marshes and ponds. Common throughout. Eu.
T. angustifolia, L. Narrow-leaved Cat-tail. Swamps. Common near salt water about New York. New Durham, N. L. Britton; Bayonne, Frank Tweedy; also at Camden, C. F. Parker; Ocean and Monmouth Cos., P. D. Knieskern. Eu.

Sparganium, Tourn. . . . . BUR-REED.
S. eurycarpum. Engelm. Closter, common, C. F. Austin.
S. simplex, Huds.; *Var.* androcladum, Engelm. Marshes and swamps. Common throughout the State.

AROIDEÆ.

Arisæma, Martius. . . . . INDIAN TURNIP.
A. triphyllum, Torr. Indian Turnip. Common in the northern and middle counties. Rare in pine barren regions. Near Hammonton, Atlantic Co., C. F. Parker; rare in Ocean Co., P. D. Knieskern; common in Monmouth Co., R. W. Brown.
A. Dracontium, Schott. Green Dragon. Rare. Low ground along Cooper's Creek, near Haddonfield, Camden Co., C. F. Parker; Sussex Co., C. F. Austin.

Peltandra, Raf. . . . . ARROW ARUM.
P. Virginica, Raf. Arrow Arum. Shallow water. Waterloo, Sussex Co., A. P. Garber; New Durham Swamp, Torrey Catalogue; and common in the middle and southern counties.

Calla, L. .   .   .   . . WATER ARUM.

C. palustris, L.   Water Arum.   New Durham Swamp, Torrey Catalogue; Budd's Lake, Morris Co., T. C. Porter; Sussex Co., C. F. Austin in Willis Catalogue.   Eu.

Symplocarpus, Salisb. .   .   . . SKUNK CABBAGE.

S. fœtidus, Salisb.   Skunk Cabbage.   Low grounds.   Common
throughout.

Orontium, L. .   . .   . GOLDEN CLUB.

O. aquaticum, L.   Golden Club.   Bergen, Torrey Catalogue; New
Durham and Closter, C. F. Austin; Budd's Lake, Morris Co., T. C.
Porter; Roseland and Springfield, Essex Co., H. H. Rusby; and com-
mon in streams on the Yellow Drift.

Acorus, L. .   .   . . .   . SWEET-FLAG.

A. Calamus, L.   Calamus.   Borders of swamps.   Sparingly in southern parts of the State, but common in the middle and northern counties.   Eu.

LEMNACEÆ.

Lemna, L. .   .   . DUCK-WEED.   DUCK'S-MEAT.

L. trisulca, L.   Closter, C. F. Austin; Passaic, W. H. Leggett; Essex
Co., H. H. Rusby.   Eu.

L. Torreyi, Austin.   In pools. Closter, C. F. Austin; near Plainfield,
Frank Tweedy.   Rare.

L. perpusilla, Torr.   Frequent in ponds.   Atlantic City, E. Diffenbaugh; Woodside, Essex Co., C. F. Austin.

Var. trinervis, Austin.   Pamrapo, Bergen Co., W. H. Leggett.

L. minor, L.   Stagnant waters.   Common throughout the State.
Eu.

Var. orbiculata, Austin.   New Durham, in ditches and marshes, C.
F. Austin.

L. polyrrhiza, L.   Ponds.   Common throughout the State.   Eu.

LILIACEÆ.

Allium, L   . . .   . . . GARLIC.   ONION.

A. tricoccum, Ait.   Wild Leek.   Rich woods; scarce.   Closter,
Bergen Co., C. F. Austin; Springfield, Essex Co., H. H. Rusby; Plainfield, Frank Tweedy; Groveville, Mercer Co., Isaac Burk.

A. Canadense, Kalm.   Wild Garlic.   Moist meadows; rather common throughout.   Camden, C. F. Parker; along the edge of the Pali-

sades, C. F. Austin; meadows near Swartswood Lake, Arthur Hollick; Passaic Falls and Bergen Point, W. H. Leggett.

A. vineale, L. Fields and pastures. Very common in the eastern counties, and frequent in other districts. Nat. Eu.

Polygonatum, Tourn. . . . . . Solomon's Seal.

P. biflorum, Ell. Smaller Solomon's Seal. Wooded banks. Quite common in the northern and middle counties and sparingly in the southern parts of the State.

P. giganteum, Dietrich. Great Solomon's Seal. Scarce. Occasional in Essex Co., H. H. Rusby; Paramus, Bergen Co., and near Troy, Morris Co., C. F. Austin; banks of the Delaware. Camden, C. F. Parker; Keyport, Monmouth Co., R. W. Brown.

Smilacina, Desf. . . . False Solomon's Seal.

S. racemosa, Desf. False Spikenard. Moist woods. Quite common throughout the State.

S. stellata, Desf. l. c. Not common. Zinc Mines, Sussex Co., C. F. Austin; Newton, Sussex Co., A. P. Garber; Snake Hill, P. V. Le Roy; Sea Breeze, Salem Co., Isaac Burk. Eu.

S. trifolia, Desf. Blue Mountains, C. F. Austin in Willis Catalogue.

Maianthemum, Weber. . . . . . Maianthemum.

M. Canadense, Desf. (Smilacina bifolia, Ker.) Moist woods. Common, especially in the northern counties.

Asparagus, L. . . . . Asparagus.

A. officinalis, L. Garden Asparagus. Common along the edges of salt marshes and by roadsides; escaped from gardens. Nat. Eu.

Lilium, Tourn. . . . . Lily.

L. Philadelphicum, L. Wild Red Lily. Quite common in the northern counties, but rare elsewhere. Keyport, Monmouth Co. (a single specimen), R. W. Brown; Chatham and Bergen Point, W. H. Leggett; Monmouth and Ocean Cos., not rare! P. D. Knieskern in Catalogue.

L. Canadense, L. Wild Yellow Lily. Moist meadows. Not common. Hoboken, Torrey Catalogue; Closter, common, C. F. Austin; Chatham, W. H. Leggett; Monmouth and Ocean Cos., P. D. Knieskern; Plainfield, Frank Tweedy.

L. superbum, L. Turk's-cap Lily. Moist meadows. Quite common throughout the State.

Erythronium, L. . . . . . Dog's-tooth Violet.
E. Americanum, Smith. Yellow Adder's Tongue. Low copses. Common throughout.
E. albidum, Nutt. White Adder's Tongue. Oxford, Hunterdon Co., F. Knighton, in Willis Catalogue; Monmouth Co. (locality now destroyed), S. Lockwood, in same.

Ornithogalum, Tourn. . . Star-of-Bethlehem.
O. umbellatum, L. Star-of-Bethlehem. Commonly escaped from gardens into moist places. Nat. Eu.

Hemerocallis, L. . . . . . . Day-lily.
H. fulva, L. Common Day-lily. Sparingly escaped from gardens into damp places or roadsides. Adv. Eu.

Uvularia, L. . . . . Bellwort.
U. perfoliata, L. Damp woods. Rather common throughout.
U. grandiflora, Smith (Includes **U.** flava, Smith). New Jersey to Virginia, rare, Gray's Manual.

Oakesia. Watson. . . . . Oakesia.
O. sessilifolia, Watson. (Uvularia, L.) Low woods. Quite common throughout the State.

Clintonia, Raf. . . . . Clintonia.
C. borealis, Raf. In a bog near Succasunna, Morris Co., T. C. Porter.

Medeola, Gronov. . . . . Indian Cucumber-root.
M. Virginiana, L. Low woods. Common in the northern and middle counties, and sparingly in the southern parts of the State.

Trillium, L. . . . . . Three-leaved Nightshade.
T. erectum, L. Birthroot. Sparingly in the northern counties. Near Norwood, Bergen Co., C. F. Austin; Preakness, Passaic Co., W. L. Fischer; Warren Co., F. Knighton; Caldwell, Essex Co., H. H. Rusby.
Var. album, Gray. Caldwell, Essex Co., rare, H. H. Rusby.
T. cernuum, L. Wake-robin. Low grounds. Quite common in the middle and northern counties.
T. erythrocarpum, Michx. Painted Trillium. In the cedar swamp at New Durham, Torrey Catalogue.

Melanthium, L. . . . . Melanthium.
M. Virginicum, L. Bunch-flower. Wet meadows. Closter, Bergen Co., C. F. Austin; Hackensack Meadows, W. H. Leggett; Tenafly,

104

PRELIMINARY CATALOGUE OF PLANTS

Bergen Co., and Green Pond, Morris Co., W. H. Rudkin; Essex Co.,
H. H. Rusby; Keyport, Monmouth Co., R. W. Brown; near Plainfield, Frank Tweedy; Camden Co., C. F. Parker; and in the southern
counties.

Veratrum, Tourn. . . . . . FALSE HELLEBORE.

V. viride, Ait. American White Hellebore. Low grounds. Quite
common in the northern and middle counties, but scarce on the Yellow Drift.

Zygadenus, Michx. . . ZYGADENE.

Z. leimanthoides, Gray. Sparingly on the Yellow Drift and confined to it. Ocean and Monmouth Cos., P. D. Knieskern; near Mount
Pleasant, Monmouth Co., R. W. Brown; near Atsion, Burlington Co.
and Tom's River, Ocean Co., C. F. Parker.

Amianthium, Gray. . . . . FLY POISON.

A. muscætoxicum, Gray. Rare. Mercer Co., Dr. Torrey in Willis
Catalogue; meadows near Camden, C. F. Parker; limestone cliffs
between Newton and Swartswood Lake, Arthur Hollick.

Helonias, L. . . . . HELONIAS.

H. bullata, L. Sparingly on the Yellow Drift formation and mostly
confined to it. Near Freehold, Monmouth Co., S. Lockwood; Manchester, Ocean Co., A. Brown; Atco, Camden Co., and near Red Bank
and Woodbury, Gloucester Co., C. F. Parker; near Colliers Mills,
Ocean Co., N. L. Britton; abundant in a bog at Succasunna, Morris
Co., (!) T. C. Porter; an interesting discovery.

Chamælirium, Willd. . . . . DEVIL'S-BIT.

C. Carolinianum, Willd. Blazing-star. Low grounds. Frequent
throughout.

Tofieldia, Huds. . . . FALSE ASPHODEL.

T. pubens, Pers. In a swamp near Manchester, Ocean Co., P. D.
Knieskern in Catalogue.

Narthecium, Moehr. . . BOG-ASPHODEL.

N. Amoricanum, Ker. Frequent in pine barren swamps on the
Yellow Drift.

Xerophyllum, Michx. . . XEROPHYLLUM.

X. setifolium, Michx. (**X.** asphodeloides, Nutt.) Near Craner's
Mills, Middlesex Co., Prof. Geo. H. Cook, and common in pine barren
regions. Not known to grow north of the Yellow Drift area.

13

SMILACEÆ.

Smilax, Tourn. . . . . . CATBRIER. GREENBRIER.

S. Walteri, Pursh. Sparingly in southern parts of the State. Cape May, C. F. Austin; Camden, W. M. Canby; near Atsion, Burlington Co., and Quaker Bridge, Atlantic Co., C. F. Parker.

S. rotundifolia, L. Greenbrier. Thickets. Common throughout.

Var. quadrangularis, Gray. Scarce. Camden, I. C. Martindale.

S. glauca, Walt. Essex Co., H. H. Rusby, and common on the Yellow Drift.

S. tamnoides, L. Thickets, New Jersey to Illinois and southward, Gray's Manual.

S. Pseudo-China, L. Dry or sandy soil. New Jersey to Kentucky and southward, Gray's Manual.

S. laurifolia, L. Sparingly in pine barren regions. Quaker Bridge, Ocean Co., C. F. Parker; southern Monmouth Co., O. R. Willis.

S. herbacea, L. Carrion Flower. Low grounds. Quite common in all parts of the State.

S. tamnifolia, Michx. Pine barrens, and confined to the Yellow Drift. Monmouth and Ocean Cos., P. D. Knieskern; near Atsion, Burlington Co., and near Camden, C. F. Parker.

JUNCEÆ.

Luzula, DC. . . . . WOOD-RUSH.

L. pilosa, Willd. Sparingly in the northern counties. Pascack, C. F. Austin; Essex Co., H. H. Rusby.

L. campestris, DC. Fields and woods. Common in the northern and middle counties, and frequent in the southern parts of the State.

Juncus, L. . . . . RUSH. BOG-RUSH.

J. effusus, L. Common Rush. Marshy grounds. Common throughout.

J. marginatus, Rostkovius. Low grounds. Quite common throughout.

Var. paucicapitatus, Engelm. Rather frequent on the Yellow Drift.

Var. biflorus, Engelm. Cape May Co., C. F. Parker.

J. bufonius, L. Low grounds. Frequent.

J. Gerardi, Loisel. Black-grass. Common on salt meadows.

J. tenuis, Willd. Low grounds. Common throughout the State.

Var. secundus, Engelm. Near Phillipsburg, T. C. Porter; Gloucester Co., C. F. Parker.

J. dichotomus, Ell. Low grounds. Frequent on the Yellow Drift and probably confined to that formation.

J. pelocarpus, E. Meyer.   Island in Lake Hopatcong, Morris Co.,
T. C. Porter; frequent on the Yellow Drift.

Var. subtilus, Engelm.   Halsey's Island in Lake Hopatcong, T. C.
Porter.

J. articulatus, L.; *Var.* obtusatus, Engelm.   In ballast on Petty's
Island, Delaware River near Camden, Isaac Burk.

J. militaris, Bigel.   Rare.   In Tom's River, Ocean Co., Atsion River,
Burlington Co., and in a pond near Atsion, C. F. Parker.   In his revi-
sion of the North American Junci, p. 461, Dr. Engelmann quotes Mr.
Parker as saying " this plant is found with submerged leaves in the
Delaware River ;" it should read, "Atsion River."   Near Manchester,
Ocean Co., P. D. Knieskern, (perhaps the same as the Tom's River
locality mentioned above).

J. acuminatus, Michx.; *Var.* debilis, Engelm.   Sparingly in the
southern counties.   Near Atsion, Burlington Co., Haleysville, Cumberland Co., and Dennisville, Cape May Co., C. F. Parker.

Var. legitimus, Engelm.   Low grounds.   Quite common throughout the State.

J. nodosus, L.   Sparingly in the northern counties.   Warren Co.,
F. Knighton; near Andover, Sussex Co., A. P. Garber.

Var. megacephalus, Torr.   Petty's Island, Delaware River near
Camden, C. F. Parker.

J. scirpoides, Lam.; *Var.* macrostemon, Engelm.   Common in the
southern and parts of the middle counties; mostly confined to the
Yellow Drift.

J. Canadensis, J. Gay.   Low grounds; not uncommon.

Var. longicaudatus, Engelm.   Frequent in the southern counties.

Var. subcaudatus, Engelm.   Red Bank, Monmouth Co., W. H. Leggett; near Camden, C. F. Parker; and probably frequent in the southern counties.

Var. coarctatus, Engelm.   Budd's Lake, T. C. Porter in Willis Catalogue.

J. asper, Engelm.   Swamps near Quaker Bridge, Atlantic Co., on
the Atsion River, and Griffith's, Camden Co., C. F. Parker.

PONTEDERIACEÆ.

Pontederia, L. . . . .   PICKEREL-WEED.

P. cordata, L.   Common in shallow water.

Var. angustifolia, Gray.   Quaker Bridge, Atlantic Co., C. F. Parker;
south end of Green Pond, Morris Co., W. H. Rudkin.

Heteranthera, R. & P. . . . . . Mud Plantain.

H. reniformis, Ruiz. & Pav. Common along the Hackensack River, and at Closter, C. F. Austin; along the Passaic River and at Chatham, W. H. Leggett; near New Brooklyn, Frank Tweedy; ditches at Camden, C. F. Parker.

Schollera, Schreb. . . . . . Water Star–grass.

S. graminea, Willd. Scarce. Whippany River, near Morristown, Eddy in Torrey Catalogue; common along the Hackensack, C. F. Austin; Swartswood Lake, H. H. Rusby; tidal mud, Delaware River at Camden, C. F. Parker.

COMMELYNEÆ.

Commelyna, Dill. . . . . Day–flower.

C. Virginica, L. Common about Camden, C. F. Parker; Fish House Station, C. & A. R. R., N. L. Britton; suburbs of Jersey City, R. W. Brown. Scarce.

Tradescantia, L. . . . . Spiderwort.

T. Virginica, L. Holland Station, Hunterdon Co., C. F. Parker; along the Delaware below Gloucester, E. Diffenbaugh; and sparingly escaped from cultivation in other parts of the State.

XYRIDEÆ.

Xyris, L. . . . . Yellow–eyed Grass.

X. flexuosa, Muhl., Chapm. Low grounds, Closter, Bergen Co., C. F. Austin; and frequent in bogs in the southern and middle counties.

X. torta, Smith. Sparingly in dry sand in pine barren regions. Near Batsto, Atlantic Co., D. C. Eaton; South Jersey, W. M. Canby in Willis Catalogue.

X. Caroliniana, Walt. Rather common in sandy swamps in the pine barrens.

X. fimbriata, Ell. Sparingly in swamps in pine barren regions, Quaker Bridge, along the Atsion River, and Atsion Meadows near Jackson, C. F. Parker; near Manchester, Ocean Co., P. D. Knieskern.

ERIOCAULONEÆ.

Eriocaulon, L. . . . . Pipewort.

E. decangulare, L. Common in swamps in the pine barrens.

E. gnaphalodes, Michx. Frequent in swamps in the pine barrens.

E. septangulare, Withering.   Edges of ponds.   Sparingly in the northern and middle counties.   Green Pond, Morris Co., W. H. Rudkin.

CYPERACEÆ.

Cyperus, L. . . . . GALINGALE.

C. flavescens, L.   Wet places.   Rare.   Long Hill and Chatham, W. H. Leggett; near Good Luck Point, Ocean Co., P. D. Knieskern; near Manchester, Ocean Co., N. L. Britton.

C. diandrus, Torr.   Low grounds.   Common throughout.

Var. castaneus, Torr.   Low grounds.   Common throughout.

C. Nuttallii, Torr.   Common in salt or brackish meadows; also in ballast at Camden, C. F. Parker.

C. erythrorhizos, Muhl.   Shore of the Delaware River, Camden, C. F. Parker.

C. inflexus, Muhl.   Petty's Island, near Camden, and at Camden, C. F. Parker; Closter, Bergen Co., 1861, C. F. Austin; near Phillipsburg, T. C. Porter.

C. compressus, L.   In ballast, Petty's Island and Camden, C. F. Parker.

C. dentatus, Torr.   Closter, Bergen Co., C. F. Austin; Franklin, Essex Co., H. H. Rusby; and common in wet places on the Yellow Drift.

C. rotundus, L.; *Var.* Hydra, Gray.   In ballast at Camden, C. F. Parker.

C. phymatodes, Muhl.   Low grounds, and occasionally appearing as a weed in cultivated fields.   Frequent.

C. strigosus, L.   Fields and low grounds.   Common throughout the State.

C. Michauxianus, Schultes.   Low grounds.   Rather rare.   Closter, Bergen Co., C, F. Austin; Manchester, Ocean Co., P. D. Knieskern in Willis Catalogue; Camden, C. F. Parker; abundant on meadows at the base of Snake Hill, N. L. Britton.

C. Engelmanni, Steud.   In ballast at Camden, C. F. Parker.

C. Grayii, Torr.   Common in sands of the sea-shore and on the Yellow Drift.

C. filiculmis, Vahl.   Common in sterile soil throughout the State.

C. Lancastriensis, T. C. Porter.   On steep banks of the Delaware River, two miles below Trenton, 1880, N. L. Britton.

C. ovularis, Torr.   Frequent in the northern and middle counties. In the southern parts of the State mostly replaced by

C. cylindricus, N. L. Britton in Bull. Torr. Club, April, 1880. Common on the Yellow Drift.

.

C. retrofractus, Torr. Sparingly on the Yellow Drift. Near Haddonfield and Griffith's, Camden Co., and Malaga, Gloucester Co., C. F. Parker; near Hoboken, (!) Dr. Torrey in Flora of N. Y.

Dulichium, Richard. . . . . DULICHIUM.
D. spathaceum, Pers. Borders of ponds and swamps. Common throughout.

Fuirena, Rottböll. . . . . . . UMBRELLA–GRASS.
F. squarrosa, Michx. Rare. Marshes at Cape May, C. F. Parker; Tom's River, Ocean Co., O. R. Willis.
Var. pumila, Torr. Near Shark River, Monmouth Co., O. R. Willis; near Long Branch, (?) C. F. Parker.

Lipocarpha, R. Br. . . . . LIPOCARPHA.
L. maculata, Torr. In ballast, Petty's Island, Camden Co., Dr. Jos. Leidy, C. F. Parker.

Hemicarpha, Nees. . . . . HEMICARPHA.
H. subsquarrosa, Nees. In ballast on Petty's Island, Camden Co., E. Diffenbaugh, Dr. Jos. Leidy.

Eleocharis, R. Br. . . . . SPIKE RUSH.
E. Robbinsii, Oakes. Rare. Monmouth Co., and Quaker Bridge, Atlantic Co., W. M. Canby; Atlantic Co., and Dennisville, Cape May Co., C. F. Parker.
E. quadrangulata, R. Br. Rare. Johnson's Pond, Dennisville, Cape May Co., T. C. Porter; Swartswood Lake, T. C. Porter; Cape May, C. F. Parker.
E. tuberculosa, R. Br. Frequent in sandy swamps on the Yellow Drift.
E. obtusa, Schultes. Muddy places. Common throughout.
E. olivacea, Torr. Wet places. Not very common. Closter, Bergen Co., C. F. Austin; abundant on Hackensack Marshes, W. H. Leggett; wet pine barrens, Ocean Co., C. F. Parker.
E. palustris, R. Br. Wet places. Common throughout.
Var. calva, Gray. Hackensack Swamps, W. H. Leggett.
E. rostellata, Torr. Marshes, Atlantic City, Cape May, and Dennisville, C. F. Parker; abundant in Hackensack Meadows, W. H. Leggett.
E. intermedia, Schultes. In swamps, N. J., Torrey Catalogue; wet banks and in swamps, Monmouth and Ocean Cos., O. R. Willis in Catalogue.

E. microcarpa, Torr.; *Var.* filiculmis, Torr.   Sparingly in pine barren regions.   Ocean Co., C. F. Austin.

E. tenuis, Schultes.   Wet places.   Quite common throughout.

E. melanocarpa, Torr.   Wet sandy places on the Yellow Drift, but no definite localities are reported.

E. tricostata, Torr.   "Near Quaker Bridge and Webb's old field, the northern limit of this plant, rare, Ocean Co.," P. D. Knieskern in Catalogue Plants Monmouth and Ocean Cos.

E. pygmæa, Torr.   Common in brackish marshes.   Occurs also at Closter, C. F. Austin. (!)

E. acicularis, R. Br.   Muddy shores.   Common throughout.

Scirpus, L. . . . . Bulrush.   Club-rush.

S. planifolius, Muhl.   Common on the Palisades, C. F. Austin; First Mt., Essex Co., H. H. Rusby.

S. subterminalis, Torr.   Quaker Bridge, tide-water in Tom's River, and in streams in Gloucester Co., C. F. Parker; Ocean Co., rare, P. D. Knieskern; Budd's Lake, T. C. Porter.

S. pungens, Vahl.   Common along the borders of ponds and streams, both fresh and brackish, in all parts of the State.

S. Olneyi, Gray.   Salt marshes.   Junction of N. Y., L. E. & W. R. R. and N. R. R. of N. J., T. F. Allen; Tom's River, Ocean Co., and Squan Village, Monmouth Co., P. D. Knieskern; Dennisville, Cape May Co., C. F. Parker.

S. validus, Vahl.   Fresh water ponds.   Rather common throughout.

S. debilis, Pursh.   Scarce.   Swamps in Monmouth and Mercer Cos., Dr. John Torrey in Willis Catalogue; not rare at Closter, C. F. Austin; Morristown, W. H. Leggett.

S. Smithii, Gray.   Budd's Lake, Morris Co., T. C. Porter; Little Timber Creek near Gloucester, A. H. Smith; tidal mud, Delaware River at Camden, C. F. Parker.

S. maritimus, L.   Sea Club-rush.   Common on salt meadows.

Var. macrostachyos, Michx.   Salt meadows.   Not as common as the type.

S. fluviatilis, Gray.   River Club-rush.   Swampy border of the Delaware River at Camden, C. F. Parker.

S. sylvaticus, L.   Closter, Bergen Co., C. F. Austin.

S. atrovirens, Muhl.   Wet meadows.   Quite common in the northern counties.

S. polyphyllus, Vahl.   Along rivulets on the Palisades, and in Somerset Co., C. F. Austin.

S. lineatus, Michx.   Scarce.   Bergen Point, W. H. Leggett; Closter, C. F. Austin.

S. Eriophorum, Michx. Wool-grass. Swamps. Common through-out the State.

Var. laxus, Gray. Bergen Point, W. H. Leggett.

Eriophorum, L. . . . . Cotton-grass.

E. Virginicum, L. Swamps. Frequent in all parts of the State, but especially abundant in pine barren regions.

E. polystachyon, L. Closter, Bergen Co., C. F. Austin; and sparingly in pine barren regions. Eu.

E. gracile, Koch; *Var.* paucinervium, Engl. Sparingly in the northern counties. Newton, Sussex Co., A. P. Garber; Budd's Lake, Morris Co., C. F. Parker; New Durham Swamp, Torrey Catalogue; Closter, C. F. Austin; Tenafly, N. L. Britton. Eu.

Fimbristylis, Vahl. . . . . . . Fimbristylis.

F. spadicea, Vahl.; *Var.* castanea, Gray. Frequent on salt meadows.

F. congesta, Torr. In ballast at Camden, C. F. Parker. Adv. Southern States.

F. autumnalis, R. & S. Low grounds. Rather common through-out.

F. capillaris, Gray. Dry sandy fields. Common throughout.

Dichromena, Richard. . . . . . . Dichromena.

D. leucocephala, Michx. Damp pine barrens of New Jersey, Gray's Manual; Monmouth and Ocean Cos., O. R. Willis. Rare.

Rhynchospora, Vahl. . . . . Beak-rush.

R. cymosa, Nutt. Rare. Near Newberry Pond, Squan, Monmouth Co., P. D. Knieskern; low ground, Griffith's, Camden Co., C. F. Parker; near Hightstown, O. R. Willis; Warren Co., F. Knighton in Willis Catalogue.

R. Torreyana, Gray. Sparingly in wet places in the pine barrens.

R. fusca, R. & S. Scarce. Abundant in swamps near Manchester, Ocean Co., P. D. Knieskern; low ground at Spring Garden, Camden Co., C. F. Parker.

R. gracilenta, Gray. Scarce and confined to the southern counties. Common (?) in Ocean Co., P. D. Knieskern; Quaker Bridge, C. F. Parker.

R. pallida, M. A. Curtis. Sparingly in pine barren regions. Bogs near Tom's River, Batsto, Atlantic Co., and Merchantville, C. F. Parker; Atsion, W. M. Canby.

R. alba, Vahl. Bogs. Common throughout.

R. Knieskernii, Carey. Sparingly on bog iron ore in the pine barrens. Paint Hollow, two miles from Manchester, Ocean Co., on the

road to Cassville, P. D. Knieskern; Quaker Bridge and near Atsion, C. F. Parker; Hope Village, Shark River, Ocean Co., A. H. Smith.

R. glomerata, Vahl. Low grounds. Rather common throughout the State. The paniculate form at Quaker Bridge, C. F. Parker.

R. cephalantha, Torr. Sparingly in sandy swamps in the southern and middle counties. Manchester, Ocean Co., P. D. Knieskern; Atsion River and at Quaker Bridge, C. F. Parker.

R. macrostachya, Torr. Manchester, Ocean Co., P. D. Knieskern; Cape May, (a glomerate form), C. F. Parker; Longacoming, Camden Co., C. E. Smith.

Cladium, P. Browne. . . . . TWIG-RUSH.

P. mariscoides, Torr. Bogs. Frequent throughout the State.

Scleria, L. . . . . NUT-RUSH.

S. triglomerata, Michx. Closter, rare, C. F. Austin; Newark Meadows, Torrey Catalogue; Chatham, W. H. Leggett; and frequent on the Yellow Drift.

S. laxa, Torr. Pine barren regions. Paint Hollow near Manchester, Ocean Co., P. D. Knieskern; banks of Mullica River near Batsto, and Tom's River, C. F. Parker.

S. pauciflora, Muhl. Near Shark River, Monmouth Co., on dry upland, rare, P. D. Knieskern.

S. verticillata, Muhl. Hackensack Meadows, 1863, T. F. Allen.

Carex, L. . . . . SEDGE.

C. polytrichoides, Muhl. Wet places. Quite common throughout.

C. Wildenovii, Schk. Rare. Bergen Co., C. F. Austin; on the Delaware River, Hunterdon Co., T. C. Porter.

C. bromoides, Schk. Swamps in the northern counties. Closter, C. F. Austin; Washington, Warren Co., A. P. Garber.

C. disticha, Huds. Sussex Co., A. P. Garber in Willis Catalogue; T. C. Porter.

C. teretiuscula, Good. Sparingly in swamps in the northern parts of the State. Common at Closter, C. F. Austin; Budd's Lake, Morris Co., T. C. Porter; Warren Co., and Andover, Sussex Co., A. P. Garber.

C. vulpinoidea, Michx. Low grounds. Common throughout.

C. stipata, Muhl. Low grounds. Common throughout.

C. cephalophora, Muhl. Woods and fields. Frequent in most districts.

C. Muhlenbergii, Schk. Dry fields. Not common. Closter, C. F. Austin; rare in Monmouth and Ocean Cos., P. D. Knieskern; Atlantic City, C. F. Parker.

C. rosea, Schk. Moist woods. Quite common throughout.

C. retroflexa, Muhl. Scarce. Closter and Palisades, C. F. Austin; near Squan Village, Monmouth Co., P. D. Knieskern.

C. tenella, Schk. Sparingly in bogs in the northern counties. New Durham, C. F. Austin; Budd's Lake, Morris Co., T. C. Porter. Eu.

C. trisperma, Dew. Swamps near Manchester, Ocean Co., rare, P. D. Knieskern; Budd's Lake, T. C. Porter; Washington, Warren Co., Spring Garden, Camden Co., Malaga, Gloucester Co., and Dennisville, Cape May Co., C. F. Parker.

C. canescens, L. Marshes and wet meadows. Rather common throughout. Eu.

C. exilis, Dew. Pine barren swamps. Manchester and Burrsville, Ocean Co., and Shark River, Monmouth Co., P. D. Knieskern; Absecom, W. M. Canby.

C. sterilis, Willd. Wet places. Quite common in the middle and northern counties.

C. stellulata, L.; Var. scirpoides, Gray. Wet places. Quite common.

Var. angustata, Gray. Closter, Bergen Co., C. F. Austin; Camden Co., C. F. Parker.

C. scoparia, Schk. Low meadows. Common throughout, and very variable.

C. lagopodioides, Schk. Low shaded places. Rather common throughout.

C. cristata, Schw. Scarce. Closter, Bergen Co., C. F. Austin.

Var. mirabilis, Boott. Plainfield, Frank Tweedy.

C. adjusta, Boott. Moist copses, New Jersey, P. D. Knieskern in Gray's Manual, p. 580; Marble Hill, above Phillipsburg, T. C. Porter.

C. fœnea, Willd. Bergen Co., C. F. Austin; Camden Co. and Atlantic City, C. F. Parker.

C. silicea, Olney (C. fœnea, Willd.; Var.? sabulonum, Boott.) Common in sands of the sea-shore.

C. straminea, Schk. Fields. Quite common and very variable.

C. alata, Torr. Atlantic City, W. M. Canby, in Willis Catalogue.

C. aquatilis, Wahl. Shores of the Delaware River north of Camden, C. F. Parker. Eu.

C. stricta, Lam. Wet meadows and swamps. Common throughout the State.

C. crinita, Lam. Wet places along streams. Common throughout.

C. littoralis, Schw. (C. Barrattii, Schw. & Torr.) Rather common in swamps in the pine barrens, Knieskern, Parker, Tweedy.

C. gynandra, Schw. Morris Co, T. C. Porter.

C. limosa, L. Peat bogs, Budd's Lake, Morris Co., C. F. Parker. Eu.

C. irrigua, Smith. Budd's Lake, Morris Co., T. C. Porter. Rare. Eu.

C. livida, Willd. Manchester, Ocean Co., P. D. Knieskern; near Atsion, Burlington Co., C. F. Parker. Eu.

C. tetanica, Schk. Newton, Sussex Co., A. P. Garber.

C. granularis, Muhl. Wet meadows. Rather common.

C. pallescens, L. Meadows, New Egypt, Ocean Co., P. D. Knieskern; Closter, C. F. Austin; Plainfield, Frank Tweedy. Scarce. Eu.

C. conoidea, Schk. Warren and Morris Cos., T. C. Porter; Closter, C. F. Austin; Paterson, W. H. Leggett; Verona, Essex Co., H. H. Rusby; Plainfield, F. Tweedy.

C. grisea, Wahl. Moist grounds. Rather common.

C. glaucoidea, Tuckm. Bergen Co., C. F. Austin; near Haddonfield, Camden Co., C. F. Parker.

C. gracillima, Schw. Bergen Co., C. F. Austin; Warren Co., C. F. Parker; Plainfield, F. Tweedy.

C. virescens, Muhl. Fields and woods. Common throughout.

C. triceps, Michx. Fields and woods. Frequent.

C. Smithii, Porter. Camden, C. F. Parker, T. C. Porter.

C. platyphylla, Carey. Near Phillipsburg, Warren Co., T. C. Porter; Delaware Water Gap, A. P. Garber; Closter and Palisades, C. F. Austin; Weehawken, M. Ruger.

C. retrocurva, Dew. Closter, Bergen Co., C. F. Austin.

C. digitalis, Willd. Frequent in the middle and northern counties.

C. laxiflora, Lam. Common throughout in one form or another.

Var. styloflexa, Boott. Closter, Bergen Co., C. F. Austin.

Var. intermedia, Boott. Shady, damp places near Camden, C. F. Parker.

Var. blanda, Dew. Hoboken, 1829, Torrey Herbarium; Closter, C. F. Austin; Plainfield, Union Co., F. Tweedy.

Var. latifolia, Boott. Palisades, C. F. Austin.

C. oligocarpa, Schk. Closter, Bergen Co., C. F. Austin.

C. eburnea, Boott. Limestone ledges, Sussex Co., C. F. Austin.

C. pedunculata, Muhl. Closter, C. F. Austin; Morristown, W. H. Leggett.

C. umbellata, Schk. Delaware Water Gap, A. P. Garber, and common in sandy fields, southern and middle counties.

C. Emmonsii, Dew. Wooded hills. Rather frequent.

C. nigromarginata, Schw. Milford, Hunterdon Co., A. P. Garber; Hartford, Burlington Co., and near Camden and Winslow, Camden Co., C. F. Parker; below Woodbury, Gloucester Co., W. M. Canby.

C. Pennsylvanica, Lam. Dry woods and hills. Common throughout.

C. varia, Muhl. Closter, C. F. Austin; Plainfield, Union Co., F. Tweedy; and probably quite common throughout.

C. miliacea, Muhl. Low grounds. Rather common.

C. scabrata, Schw. Closter and Palisades, C. F. Austin; Warren Co., T. C. Porter; near Andover, Sussex Co., A. P. Garber.

C. arctata, Boott. Closter, C. F. Austin in Bull. Torr. Bot. Club VI.,11.

C. glabra, Boott. In a sphagnous swamp, six miles southeast of Camden, and at Absecom, W. M. Canby; East Creek, Cape May Co., C. F. Parker.

C. debilis, Michx. Moist meadows. Rather frequent throughout the State. Closter, C. F. Austin; Bergen Point and Chatham, W. H. Leggett; Plainfield, Frank Tweedy; Atlantic City and Camden, C. F. Parker.

C. flava, L. Newton, Sussex Co., A. P. Garber; also in ballast (?) at Kaign's Point, Camden, 1865, C. F. Parker. Eu.

C. filiformis, L. Sparingly in the northern counties. Abundant along the Buckman road one-half mile northeast of Closter, C. F. Austin; Budd's Lake, C. F. Parker. Eu.

C. lanuginosa, Michx. Near Squan Village, Monmouth Co., P. D. Knieskern; Closter, C. F. Austin; near Washington, Warren Co., and in Camden Co., C. F. Parker; Essex Co., H. H. Rusby.

C. vestita, Willd. Closter, C. F. Austin; Washington, Warren Co., A. P. Garber; Franklin, Essex Co., H. H. Rusby; Plainfield, Frank Tweedy; and common in sandy fields in the southern and middle counties.

C. polymorpha, Muhl. Rare in Monmouth and Ocean Cos., P. D. Knieskern; near Washington, Warren Co., A. P. Garber; Plainfield, Frank Tweedy.

C. striata, Michx. Quite common in pine barren regions, and mostly confined to the Yellow Drift.

C. riparia, Curtis. Frequent. Closter, C. F. Austin; near Andover, Sussex Co., A. P. Garber; bogs at Manchester, Ocean Co., Frank Tweedy; Budd's Lake, T. C. Porter; river swamps along the Delaware in Gloucester Co., C. F. Parker. Eu.

C. trichocarpa, Muhl. Closter, C. F. Austin in Bull. Torr. Bot. Club, VI., 11.

C. comosa, Boott. Wet places. Quite common in the northern and middle counties, but sparingly on the Yellow Drift.

C. Pseudo-Cyperus, L. Closter, Bergen Co., C. F. Austin. Eu.

C. hystricina, Willd. Wet meadows. Frequent. Closter, C. F. Austin, Camden, C. F. Parker, Ocean and Monmouth Cos., P. D. Knieskern.

C. tentaculata, Muhl. Wet meadows. Common throughout the State.

C. intumescens, Rudge. Wet meadows. Common throughout the State.

C. Grayii, Carey. Closter, C. F. Austin in Bull. Torr. Bot. Club, VI. 12.

C. lupulina, Muhl. Closter, C. F. Austin; Cape May, C. F. Parker; South Jersey, W. M. Canby; Roseland, Essex Co., H. H. Rusby.

C. lupuliformis, Sartwell. Hackensack Flats, in woods, C. F. Austin, in Bull. Torr. Bot. Club, VI. 12.

C. folliculata, L. Low grounds. Rather common in all parts of the State.

C. subulata, Michx. Cedar Swamp, Weehawken, Torrey Catalogue ; New Durham, R. H. Brownne; Malaga, Gloucester Co., and Camden Co., C. F. Parker; South Amboy, W. H. Leggett; Ocean and Monmouth Cos., P. D. Knieskern.

C. squarrosa, L. Low grounds. Rather common except in the pine barrens.

C. utriculata, Boott. Hackensack swamps and along N. R. R. of N. J., between Bergen and New Durham, W. H. Leggett; Budd's Lake, C. F. Parker; Camden, T. C. Porter, in Willis Catalogue.

C. Schweinitzii, Dewey. Wet swamps, New Jersey, Gray's Manual, p. 600.

C. monile, Tuck. Closter, Bergen Co., C. F. Austin; Franklin, Sussex Co., A. P. Garber. Scarce.

C. Tuckermani, Boott. English Neighborhood, Bergen Co., C. F. Austin in herb. C. F. Parker.

C. bullata, Schk. Closter, C. F. Austin; Atlantic and Camden Cos., C. F. Parker; not rare in Ocean and Monmouth Cos., P. D. Knieskern.

GRAMINEÆ.

Leersia, Solander. .   . .   . WHITE GRASS.

L. Virginica, Willd. White Grass. Damp shady places. Quite common throughout.

L. oryzoides, Swartz. Rice Cut-grass. Wet places. Common throughout. Eu.

Zizania, Gronov. . . . . INDIAN RICE.

Z. aquatica, L. Indian Rice. Wild Oats. Swamps along rivers and streams. Common in most districts.

Alopecurus, L. . . . . FOXTAIL GRASS.

A. geniculatus, L. Floating Foxtail. In ballast at Camden, C. F. Parker. Adv. Eu.

A. aristulatus, Michx. Wild Foxtail. Closter, C. F. Austin; Newton, Sussex Co., A. P. Garber; Bergen Point and Palisades, W. H.

Leggett; Mercer Co., Dr. John Torrey; Camden, and river swamps in Gloucester Co., C. F. Parker. Eu.

Phleum, L. . . . . CAT'S-TAIL GRASS.

P. pratense, L. Timothy. Herd's Grass. Fields and meadows. Common throughout the State. Nat. Eu.

Crypsis, Ait. . . . . CRYPSIS.

C. echœnoides, Lam. Waste places and ballast at Camden, C. F. Parker. Nat. Eu.

Vilfa, Adans., Beauv.   . . . . RUSH-GRASS.

V. aspera, Beauv. Sparingly throughout the State. Closter, C. F. Austin; Carpentersville, A. P. Garber; Ocean and Monmouth Cos., P. D. Knieskern.

V. vaginæflora, Torr. Dry fields. Quite common throughout.

Sporobolus, R. Br. . . . . . . DROP-SEED GRASS.

S. compressus, Kunth. Frequent in bogs in the pine barrens.

S. serotinus, Gray. Common in sandy swamps on the Yellow Drift, and mostly confined to that formation.

Agrostis, L. . . . . BENT-GRASS.

A. elata, Trin. Frequent in pine barren swamps.

A. perennans, Tuck. Thin-grass. Damp shaded places. Rather common.

A. scabra, Willd. Hair-grass. Dry or damp open places. Rather common.

A. canina, L. Brown Bent-grass. Near Andover, T. C. Porter in Willis Catalogue. Adv. Eu. (?)

A. vulgaris, With. Red-top. Herd's Grass. Low meadows; commonly cultivated, and naturalized from Europe. Probably not indigenous in New Jersey. Eu.

A. alba, L. Fiorin. White Bent-grass. Meadows and fields; introduced from Europe for a pasture grass. Not indigenous in New Jersey. Eu.

Polypogon, Desf.   . . . . BEARD-GRASS.

P. Monspeliensis, Desf. Beard-grass. In ballast at Camden, C. F. Parker. Adv. Eu.

Cinna, L. . . . . WOOD REED-GRASS.

C. arundinacea, L. Wood Reed-grass. Moist woods. Rather common in the northern and middle counties, and frequent on the Yellow Drift. Eu.

Muhlenbergia, Schreb. . . . . Drop-seed Grass.

M. sobolifera, Trin. Rocky woods in the northern and middle counties. Closter, C. F. Austin; Snake Hill, N. L. Britton; Verona, Essex Co., H. H. Rusby.

M. glomerata, Trin. Bogs in the northern counties. Closter, C. F. Austin.

M. Mexicana, Trin. Low grounds. Quite common throughout the State.

M. sylvatica, Torr. & Gray. Low shaded places, northern and middle counties. Not common.

M. Willdenovii, Trin. Closter, Bergen Co., C. F. Austin; not common in Monmouth and Ocean Cos., P. D. Knieskern.

M. diffusa, Schreb. Nimble Will. Dry hills and woods. Rather common.

M. capillaris, Kunth. Hair Grass. Snake Hill and Little Snake Hill, W. H. Leggett; N. L. Britton, 1880; "sandy soils, south, very rare," W. M. Canby in Willis Catalogue.

Brachyelytrum, Beauv. . . . . Brachyelytrum.

B. aristatum, Beauv. Rare. Near Shark River, Monmouth Co., P. D. Knieskern; Camden Co., C. F. Parker; Long Hill, W. H. Leggett.

Calamagrostis, Adans. . . . Reed Bent-grass.

C. Canadensis, Beauv. Blue Joint-Grass. Common in low meadows near Squan and Shark Rivers, Monmouth Cos., P. D. Knieskern; and common in the northern counties.

C. Nuttalliana, Steud. Scarce. Palisades and Closter, C. F. Austin; Newton, Sussex Co., A. P. Garber; rare in Monmouth and Ocean Cos., P. D. Knieskern; Camden Co., C. F. Parker; Morristown and Secaucus, W. H. Leggett.

C. brevipilis, Gray. Sparingly in pine barren swamps. Burlington and Atlantic Cos., C. F. Parker; near Manchester, Ocean Co., N. L. Britton.

C. arenaria, Roth. Sea Sand-reed. Common on sands of the seashore. Eu.

Oryzopsis, Michx. . . . . Mountain Rice.

O. melanocarpa, Muhl. Sparingly in the northern parts of the State. Bergen Co., C. F. Austin; on First Mt., Essex Co., H. H. Rusby; Franklin, Sussex Co., A. P. Garber.

O. asperifolia, Michx. Woods, Plainfield, Frank Tweedy. Scarce.

O. Canadensis, Torr. Rare. Essex Co., H. H. Rusby, W. M. Wolfe; east side of Swartswood Lake, Morris Co., Arthur Hollick.

Stipa, L. . . . . FEATHER-GRASS.

S. avenacea, L.   Black Oat-grass.   Palisades, C. F. Austin ; Closter, W. H. Leggett; and quite common in woods on the Yellow Drift.

Aristida, L. . . . . TRIPLE-AWNED GRASS.

A. dichotoma, Michx.   Poverty Grass.   Dry sandy fields.   Common throughout.

A. gracilis, Ell.   Dry sandy fields.   Quite common throughout.

A. purpurascens, Poir.   Limestone rocks, Sussex Co., C. F. Austin ; Carpentersville, A. P. Garber; not rare in Ocean and Monmouth Cos., P. D. Knieskern ; and frequent in the southern counties.

A. tuberculosa, Nutt.   Middletown, Monmouth Co., P. D. Knieskern ; Sandy Hook, M. Ruger; South Amboy, N. L. Britton.

Spartina, Schreb. . . . . MARSH GRASS.

S. cynosurioides, Willd.   Hackensack Meadows, W. H. Leggett and T. F. Allen.

S. polystachya, Willd., Muhl.   Salt Reed-grass.   Common along salt marshes and salt water ditches.

S. juncea, Willd.   Salt Marsh Grass.   Common on salt meadows.   Eu.

S. stricta, Roth. ; *Var.* glabra, Gray.   Salt Marsh Grass.   Common in ditches on salt meadows.   Eu.

Var. alternifolia, Gray.   Cape May, W. M. Canby in Willis Cata. logue.

Bouteloua, Lagasca. . . . . MUSKIT-GRASS.

B. curtipendula, Gray.   Limestone ledges, Sussex Co., C. F. Austin ; Newton, Sussex Co., A. P. Garber.   Scarce.

Gymnopogon, Beauv. . . . . . NAKED-BEARD GRASS.

G. racemosus, Beauv.   Sparingly on the Yellow Drift.   South Jersey, W. M. Canby; Griffith's, Camden Co., C. F. Parker.

Cynodon, Richard. . . . . . . BERMUDA GRASS.

C. Dactylon, Pers.   Frequent in waste ground and ballast at Camden, C. F. Parker; and in ballast at Jersey City, Addison Brown. Adv. Eu.

Dactyloctenium, Willd. . . . . . EGYPTIAN GRASS.

D. Ægyptiacum, Willd.   Egyptian Grass.   In ballast at Camden, C. F. Parker; and at Jersey City, A. Brown.   Adv. Africa.

Eleusine, Gærtn. . . . . YARD-GRASS.

E. Indica, Gærtn.   Wire-grass.   Cultivated fields and roadsides. Very common throughout.   Nat. India.

Leptochloa, Beauv. . . . . . . LEPTOCHLOA.
L. fascicularis, Gray. Frequent along the edges of salt marshes.

Tricuspis, Beauv. . . . . TRICUSPIS.
T. seslerioides, Torr. Tall Red-top. Dry sandy fields. Rather common throughout.
T. purpurea, Gray. Sand Grass. Common in sands of the sea shore and in dry pine barrens.

Dactylis, L. . . . . . . ORCHARD GRASS.
D. glomerata, L. Orchard Grass. Common in fields and meadows. Nat. Eu.

Eatonia, Raf. . . . . EATONIA.
E. obtusata, Gray. Passaic Falls and Weehawken, W. H. Leggett; and probably more common than hitherto supposed.
E. Pennsylvanica, Gray. Moist woods. Rather common in all parts of the State.

Glyceria, R. Br., Trin. . . . . . . MAUNA-GRASS.
G. Canadensis, Trin. Rattlesnake-grass. Wet places. Scarce in Monmouth and Ocean Cos., P. D. Knieskern; Camden Co. and Closter, Bergen Co., C. F. Parker; Plainfield, Frank Tweedy; New Durham, N. L. Britton; Parsippany, H. H. Rusby.
G. obtusa, Trin. Common in the pine barren regions, and mostly confined to the Yellow Drift. Homestead Station, N. R. R. of N. J., M. Ruger.
G. elongata, Trin. Budd's Lake, Morris Co., T. C. Porter in Willis Catalogue; Carlstadt, Hudson Co., W. H. Leggett.
G. nervata, Trin. Wet meadows. Common throughout the State.
G. pallida, Trin. Shallow water. Quite common throughout.
G. fluitans, R. Br. Shallow water. Rather common throughout the State. Eu.
G. acutiflora, Torr. Bergen Co., C. F. Austin; near Waterford, Merchantville and Gloucester, Camden Co., C. F. Parker; Hoboken, W. H. Leggett.
G. distans, Wahl. Abundant in ballast at Camden, C. F. Parker; and Communipaw, J. Schrenck. Eu.

Brizopyrum, Link. . . . . SPIKE-GRASS.
B. spicatum, Hook. Spike-grass. Common in salt marshes.

Poa, L. . . . . . . MEADOW-GRASS. SPEAR-GRASS.
P. annua, L. Low Spear-grass. Very common in waste and cultivated grounds; probably not indigenous to New Jersey. Nat. Eu.

15

P. compressa, L. Wire-grass. Dry fields. Rather common throughout the State; probably not native to New Jersey. Nat. Eu.

P. serotina, Ehrh. False Red-top. Camden, C. F. Parker; Closter, C. F. Austin; Roseland, Essex Co., H. H. Rusby; Bergen Point, W. H. Leggett. Eu.

P. pratensis, L. Kentucky Blue-grass. Common throughout, but introduced for a pasture grass and not native to New Jersey. Nat. Eu.

P. trivialis, L. Roughish Meadow-grass. Not common. Meadows, Camden, C. F. Parker; Monmouth Co., O. R. Willis; New Durham, N. L. Britton; Bergen Point, W. H. Leggett. Nat. Eu.

Eragrostis, Beauv. . . . . ERAGROSTIS.

E. reptans, Nees. Shore of Delaware River, above Phillipsburg, T. C. Porter; Petty's Island, near Camden, C. F. Parker.

E. poæoides, Beauv.; *Var.* megastachya, Gray. Sandy waste places. Rather common throughout. Nat. Eu.

E. pilosa, Beauv. Waste ground and ballast at Camden, C. F. Parker; not rare in Monmouth and Ocean Cos., P. D. Knieskern. Nat. Eu.

E. Frankii, Meyer. Shore of the Delaware River, above Phillipsburg, T. C. Porter.

E. Purshii, Schrader. Sandy soil. Shore of the Delaware, above Phillipsburg, T. C. Porter; and common in the middle and southern counties.

E. capillaris, Nees. Sandy fields of Ocean and Monmouth Cos., not common, P. D. Knieskern; Gloucester and Cumberland Cos., C. F. Parker; Little Snake Hill, N. L. Britton; near Phillipsburg, T. C. Porter.

E. pectinacea, Gray. Dry fields. Rather common throughout.

Festuca, L. . . . . FESCUE–GRASS.

F. Myurus, L. Scarce. Squan, Monmouth Co., W. H. Leggett, 1857; Atco, Camden Co., J. H. Redfield; Camden, C. E. Smith; and in ballast at Camden, C. F. Parker. Nat. Eu.

F. tenella, Willd. Frequent in sandy fields, southern and middle counties, and sparingly in the northern parts of the State. Essex Co., H. H. Rusby.

F. ovina, L.; *Var.* duriuscula, Gray. Fields, &c. Rather common throughout. Nat. Eu.

F. elatior, L.; *Var.* pratensis, Gray. Meadow-fescue. Meadows. Common throughout. Nat. Eu.

F. nutans, Willd. Scarce. Palisades, C. F. Austin; Bergen Point, W. H. Leggett.

Bromus, L. . . . .   BROME–GRASS.

B. secalinus, L.   Cheat.   Chess.   Very common in wheat fields.
Adv. Eu.

B. racemosus, L.   Upright Chess.   Fields and meadows.   Common.
Adv. Eu.

B. mollis, L.   Soft Chess.   In ballast at Communipaw, N. L. Britton.   Adv. Eu.

B. ciliatus, L.   Hairy Chess.   Woodlands.   Rather common in
the northern and middle counties.

B. sterilis, L.   Waste places and ballast at Camden, C. F. Parker;
Bergen Point, W. H. Leggett; Passaic, C. F. Austin; Plainfield, Frank
Tweedy; common in New Brunswick, N. L. Britton; Trenton, I. S.
Moyer; ballast at Communipaw, A. Brown.   Nat. Eu.

Uniola, L. . . . .   SPIKE–GRASS.

U. gracilis, Michx.   Common in the sands of the Yellow Drift, and
mostly confined to that formation.

Phragmites, Trin. . . . .   REED.

P. communis, Trin.   Swamps and edges of ponds.   Frequent.   Newton, Sussex Co., A. P. Garber; near Cape May, C. F. Parker; Good
Luck Meadows, Ocean Co., P. D. Knieskern; Budd's Lake, T. C.
Porter; common on Newark and Hackensack Meadows, N. L. Britton.   Eu.

Lolium, L. . . . .   DARNEL.

L. perenne, L.   Common Darnel.   Fields and roadsides.   Quite
common.   Nat. Eu.

L. temulentum, L.   Bearded Darnel.   In ballast at Camden, C. F.
Parker.   Adv. Eu.

Triticum, L. . . . .   WHEAT.

T. repens, L.   Quitch-grass.   Common in cultivated fields and along
roadsides.   Nat. Eu.

Hordeum, L. . . . .   BARLEY.

H. jubatum, L.   Squirrel-tail Grass.   In ballast at Camden, C. F.
Parker.   Adv. Lake Superior. (?)

Elymus, L. . . . . . .   LYME–GRASS.   WILD RYE.

E. Virginicus, L.   Wild Rye.   Banks of streams and rivers.   Quite
common throughout.

E. Canadensis, L.   Nodding Wild Rye.   Scarce.   Banks of Shark
River, Monmouth Co., P. D. Knieskern; Sussex Co., and Palisades, C.
F. Austin; Hackensack meadows, W. H. Leggett.

E. striatus, Willd.   Sparingly in woodlands, northern and middle counties, Franklin, Essex Co., H. H. Rusby.

Gymnostichum, Schreb. . . BOTTLE-BRUSH GRASS.
G. Hystrix, Schreb. Bottle-brush Grass. Palisades, C. F. Austin; Warren Co., C. F. Parker; Weehawken, M. Ruger.

Danthonia, DC. . . . . WILD OAT-GRASS.
D. spicata, Beauv. Dry sterile soil. Common throughout the State.
D. sericea, Nutt. Sandy soil. Frequent on the Yellow Drift.

Avena, L. . . . . OAT.
A. striata, Michx. Wild Oat. Rocky woods, Palisades, C. F. Austin, in Willis Catalogue.

Trisetum, Pers. . . . . TRISETUM.
T. palustre, Torr. Scarce. Meadows at Closter, C. F. Austin; Washington, Warren Co., A. P. Garber.

Aira, L. . . . . HAIR-GRASS.
A. flexuosa, L. Common Hair-grass. Dry sandy woods. Quite common throughout. Eu.
A. cæspitosa, L. "Damp places, rare," P. D Knieskern, in Catalogue of Plants of Ocean and Monmouth Cos. Eu.
A. præcox, L. Sandy fields, Camden, J. H. Redfield; near Gloucester, C. F. Parker. Nat. Eu.
A. caryophyllea, L. Roadsides near Salem, W. M. Canby in Willis Catalogue. Nat. Eu.

Arrhenatherum, Beauv. . . . . . . OAT-GRASS.
A. avenaceum, Beauv. Grass of the Andes. Closter, C. F. Austin; Shark River, Monmouth Co., P. D. Knieskern. Nat. Eu.

Holcus, L. . . . . MEADOW SOFT-GRASS.
H. lanatus, L. Velvet-grass. Meadows. Quite common throughout the State. Nat. Eu.

Hierochloa, Gmelin. . . . . HOLY GRASS.
H. borealis, R. & S. Vanilla Grass. Border of salt marshes, near Squan Village, Monmouth Co., rare, P. D. Knieskern; Newark Meadows, N. L. Britton; Salem, W. M. Canby. Eu.

Anthoxanthum, L. . . . . . . VERNAL GRASS.
A. odoratum, L. Sweet Vernal Grass. Fields and pastures. Quite common throughout. Nat. Eu.

Phalaris, L. . . . . CANARY-GRASS.

P. Canariensis, L. Canary grass. Closter, C. F. Austin, in Willis Catalogue; waste places and ballast at Camden, C. F. Parker; ballast at Communipaw, A. Brown; Hoboken, W. H. Leggett. Adv. Eu.

P. arundinacea, L. Reed Canary-grass. Wet places. Frequent in the northern counties. Also in ballast at Camden, C. F. Parker. Eu.

Amphicarpum, Kunth. . . . . . AMPHICARPUM.

A. Purshii, Kunth. Abundant in pine barren regions, and confined to the Yellow Drift.

Paspalum, L. . . . . PASPALUM.

P. Walterianum, Schultes. Cape May, Nuttall in Gray's Manual, p. 645.

P. setaceum, Michx. Sandy fields. Rather common throughout.

P. læve, Michx. Ocean and Monmouth Cos., P. D. Knieskern; Camden and Cape May, C. F. Parker; Franklin, Essex Co., H. H. Rusby.

P. distichum, L. Joint Grass. Quite abundant in damp ballast at Camden, C. F. Parker. Adv. Southern States.

Panicum, L. . . . . PANIC-GRASS.

P. filiforme, L. Slender Crab-grass. Sandy fields. Common in the southern and middle counties, and sparingly in the northern parts of the State.

P. glabrum, Gaudin. Smooth Crab-grass. Waste and cultivated fields. Frequent. Nat. Eu.

P. sanguinale, L. Common Crab-grass. Waste and cultivated grounds. Very common. Nat. Eu.

P. anceps, Michx. Frequent in pine barren swamps.

P. agrostoides, Spreng. Wet meadows and shores. Quite common throughout.

P. proliferum, Lam. Wet places. Abundant along the edges of salt marshes, and sparingly along fresh water swamps. Closter, C. F. Austin.

P. capillare, L. Odd-witch Grass. Fields and roadsides. Very common throughout.

P. virgatum, L. Moist sandy soil. Common throughout.

P. amarum, Ell. Cape May Point near the Inlet, C. F. Parker; Sandy Hook, M. Ruger.

P. latifolium, L. Moist thickets. Common throughout.

P. clandestinum, L. Moist thickets. Rather common throughout.

P. viscidum, Ell. Damp ground. Camden, and Dennisville, Cape May Co., C. F. Parker. Rare, and confined to the southern counties.

P. pauciflorum, Ell. Wet meadows and copses. Rather common in all parts of the State.

P. dichotomum, L. Everywhere. Very common throughout and extremely variable.

P. depauperatum, Muhl. Dry woods and hills. Quite common.

P. verrucosum, Muhl. Frequent in pine barren swamps and mostly confined to the Yellow Drift.

P. Crus-galli, L. Barnyard-grass. Waste places. Common. Nat. Eu. *Var.* hispidum, Gray. Common in salt or brackish marshes.

Setaria, Beauv. . . . . . BRISTLY FOX-TAIL GRASS.

S. verticillata, Beauv. Whorled Foxtail. Cultivated fields, Ocean and Monmouth Cos., not common, P. D. Knieskern ; Closter, Bergen Co., C. F. Austin ; Newark and Hoboken, W. H. Leggett. Nat. Eu.

S. glauca, Beauv. Foxtail. Cultivated and waste grounds. Very common throughout. Nat. Eu.

S. viridis, Beauv. Green Foxtail. Cultivated grounds. Frequent. Adv. Eu.

S. Italica, Kunth. Millet. Bengal Grass. Near Coopers Creek, Camden, 1873, C. F. Parker ; (not found there since.)

Cenchrus, L. . . . . BURR-GRASS.

C. tribuloides, L. Common Burr-grass. Common on sands of the sea-shore and throughout the Yellow Drift area. Also Delaware Water Gap, H. H. Rusby.

Tripsacum, L. .   . . . GAMA-GRASS.

T. dactyloides, L. Sesame-grass. Scarce. Border of pond four miles north of Egg Harbor City, C. F. Parker ; Monmouth Beach Centre, Addison Brown.

Erianthus, Michx. . . . . WOOLLY BEARD-GRASS.

E. alopecuroides, Ell. Sparingly in pine barren regions. Near Hammonton, Atlantic Co., and Camden Co., C. F. Parker.

Andropogon, L. . . . . BEARD-GRASS.

A. furcatus, Muhl. Morris Co., C. F. Austin ; rare in Ocean and Monmouth Cos., P. D. Knieskern ; Verona, Essex Co., H. H. Rusby ; Little Snake Hill, N. L. Britton.

A. scoparius, Michx. Common in dry fields throughout the State.

A. Virginicus, L. Dry sandy soil. Rather common, except in the northern counties.

A. macrourus, Michx. Common in the sands of the Yellow Drift.

Sorghum, Pers. . . . . BROOM CORN.

S. nutans, Gray. Indian Grass. Dry soil. Common throughout the State.

APPENDIX I.

Species of Phanerogamous plants hitherto published as growing wild in New Jersey, but now considered as not sufficiently
authenticated to be admitted into this Catalogue without further
identification.

Stellaria pubera, Michx. "Middle and southern, not rare," O. R.
Willis in Catalogue.

Holosteum unbellatum, L. "Morris Co.," C. F. Austin in Willis
Catalogue. The plant here referred to is certainly not a Holosteum; Mr. C. F. Parker has it in his herbarium, and it is not
determined.

Æschynome hispida, Willd. "Banks of the Delaware below
Kaighn's Ferry, Camden, very rare," Barton in Com. Flo. Phil.,
II., p. 30. No trace of this plant exists now (C. F. Parker).

Polymnia Canadensis, L. "Weehawken," Gray's Manual, p. 248,
and Willis Catalogue, has been shown to be **P.** Uvedalia, L. See
Bull. Torr. Bot. Club, I., 4.

Cirsium altissimum, Spreng. "Fields and copses, common," Knieskern, Catalogue of Plants of Monmouth and Ocean Cos., is certainly a mistake.

Cirsium Virginianum, Michx. "Open grounds, Monmouth Co.,"
Willis Catalogue.

Mulgedium Floridanum, DC. "Common in Monmouth Co.," Willis Catalogue.

Calluna vulgaris, Salisb. Reported by Dr. O. R. Willis as growing
wild near Egg Harbor, but shown by Mr. Thomas Meehan to
have been planted there. See Bull. Torr. Bot. Club, VI., 252, 265.

Veronica spicata, L. "Escaped from gardens, north," Willis Catalogue. This is a cultivated European species, and is not regarded
as deserving a place in this Catalogue.

Gerardia integrifolia, Gray. Austin in Willis Catalogue, but no locality given.

Salvia urticifolia, L. "Mountains," Beck in Willis Catalogue. Certainly erroneous, as the plant has not been found north of Maryland.

Polemonium reptans, L. "Belvidere," Knighton in Willis Catalogue.

Gentiana alba, Muhl. "Hunterdon Co.," Knighton in Willis Catalogue.

Polygonum cilinode, Michx. "Mountains, north," Willis Catalogue, but no locality given.

Fraxinus viridis, Michx. "Near streams, not rare," P. D. Knieskern in Catalogue of Plants of Monmouth and Ocean counties.

Salix petiolaris, Smith. "Warren Co.," F. Knighton in Willis Catalogue.

Salix viminalis, L. "Warren Co.," F. Knighton in Willis Catalogue.

Allium cernuum, Roth. "Rocky banks," Willis Catalogue, is probably a mistake.

Juncus Roemerianus, Scheele. "Brackish marshes, New Jersey," Pursh, in Gray's Manual, is doubted.

Carex sparganioides, Muhl. "Common," Willis Catalogue. No localities are reported for this plant.

Carex stellulata, L. "Wet meadows and marshes, common," Willis Catalogue. Is not known to grow nearly so far south.

Carex salina, Wahl. "On banks of a branch of Tom's River; this species is included with some hesitation," Knieskern, Catalogue of Plants of Monmouth and Ocean counties.

APPENDIX II.

List of Plants, mostly of European origin, found on ballast deposits at Camden and in the vicinity of New York. Where not otherwise stated, the plants from Camden were collected by Mr. C. F. Parker, and those from Communipaw and Hoboken by Mr. Addison Brown. These species are not mentioned in Gray's Manual of Botany.

Ranunculus philonotis, Ehrh. Camden.
Ranunculus arvensis, L. Communipaw.
Ranunculus lanuginosus, L. Communipaw.
Papaver Rhoeas, L. Camden, Communipaw.
Fumaria capreolata, L. Camden.
Escholtzia Californica, Cham. Communipaw.
Diplotaxis tenuifolia, DC. Camden, Communipaw, Hoboken.
Diplotaxis muralis, DC. Camden, Isaac Burk.
Diplotaxis virgata, DC. Camden, I. C. Martindale.
Diplotaxis ericoides, DC. Communipaw.
Brassica cheiranthus, Vill. Hoboken.
Erysimum orientale, L. Camden.
Erysimum repandum, L. Communipaw.
Sisymbrium Irio, L. Camden.
Lepidium graminifolium, L. Camden, Communipaw, Hoboken.
Lepidium Smithii, Hook. Camden, Communipaw, I. C. Martindale.
Rapistrum rugosum, All. Communipaw.
Iberis umbellata, L. Communipaw, J. Schrenck.
Neslia paniculata, Desv. Communipaw.
Polanisia viscosa, DC. Communipaw, Camden.
Reseda lutea, L Communipaw, Camden.
Reseda odorata, L. Communipaw, Camden.
Reseda alba, L. Camden.
Ionidium parviflorum, Vent. (?) Communipaw.
Frankenia pulverulenta, L. Communipaw.
Lychnis chalcedonica, L. Communipaw.
Lychnis diurna, Sibth. Camden, Communipaw.
Lychnis Flos-cuculi, L. Communipaw.

Silene Gallica, L.   Camden, Communipaw. (?)
Silene pendula, L.   Communipaw.
Polycarpon tetraphyllum, L.   Camden.
Corrigiola litoralis, L.   Camden, Communipaw.
Portulaca pilosa, L.   Camden, Communipaw.
Gossypium Barbadense, L.   Communipaw.
Gossypium album, Ham.   Camden.
Sida rhombifolia, L.   Communipaw, Camden.
Sida carpinifolia, L.; *Var.* brevicuspidata, Gris.   Communipaw.
Sida stipulata, Chap.   Camden.
Pavonia hastata, Cav.   Communipaw.
Waltheria Americana, L.   Communipaw.
Malvastrum tricuspidatum, Gray.   Camden, Communipaw.
Malva borealis, Wallm.   Camden.
Spheralcea miniata, Cav.   Camden.
Abutilon cordifolia, L.   Camden.
Corchorus fascicularis, Lam.   Camden.
Linum angustifolium, Huds.   Camden.
Tribulus terrestris, L.   Camden, Communipaw.
Geranium rotundifolium, L.   Communipaw.
Geranium molle, L.   Camden.
Erodium moschatum, L'Her.   Communipaw, Camden.
Oxalis corniculata, L.   Communipaw.
Ulex nanus, Forst.   Communipaw.
Ononis antiquorum, L.   Camden, I. C. Martindale.
Ononis spinosa, L.   Hoboken.
Cardiospermum halicacabum, L.   Camden.
Anthyllis vulneraria, L.   Communipaw.
Lupinus angustifolius, L.   Camden.
Melilotus arvensis, Wallr.   Communipaw.
Melilotus parviflora, Desf.   Camden, Communipaw.
Melilotus gracilis, DC.   Communipaw.
Melilotus sulcata, Desf.   Communipaw, Camden.
Melilotus macrorhiza, Pers.   Camden.
Melilotus compacta, Salz.   Camden.
Medicago marginata, Willd.   Communipaw, Camden.
Medicago falcata, L.   Hoboken, I. C. Martindale.
Medicago minima, Lam.   Communipaw, Camden.
Medicago apiculata, Willd.   Camden, I. C. Martindale, C. A. Boice.
Medicago pubescens, DC.   Camden.
Medicago littoralis, Rohdl.   Camden.
Medicago echinus, Willd.   Camden.
Medicago spinulosus, DC.   Camden, I. C. Martindale.

Medicago turbinata, Willd.   Camden.
Medicago muricata, All.   Camden.
Trifolium elegans, Reich.   Communipaw.
Trifolium Carolinianum, Michx.   Camden.
Trifolium maritimum, Huds.   Camden.
Trifolium hybridum, L.   Camden.
Trifolium lappaceum, L.   Camden.
Trifolium subterraneum, L.   Camden, I. C. Martindale.
Dorycnium hirsutum, DC.   Communipaw.
Sasbania Floridana, Wats.   Camden.
Ornithopus compressus, L.   Communipaw.
Ornithopus perpusillus, L.   Camden.
Astragalus glycyphyllos, L.   Communipaw.
Adesmia muricata, DC.; *Var.* dentata. (?)   Communipaw.
Vignea luteola, Benth.   Camden, Communipaw.
Arachis hypogæa, Willd.   Camden, Communipaw.
Ervum Lens, L.   Communipaw, M. Ruger; Camden.
Lathyrus sativus, L.   Communipaw.
Lathyrus Aphaca, L.   Communipaw, Camden, I. C. Martindale.
Lathyrus scorpiurus.   Camden.
Phaseolus ochrus, L.   Camden.
Cassia occidentalis, L.   Communipaw.
Rhynchosia minima, DC.   Communipaw.
Desmanthus brachylobus, Benth.   Communipaw.
Poterium Sanguisorba, L.   Communipaw.
Potentilla reptans, L.   Camden.
Ammania latifolia, L.   Camden.
Epilobium hirsutum, L.   Communipaw.
Epilobium pubescens, Roth.   Camden.
Ecballium agreste, Reich.   Communipaw, Camden.
Trianthema monogynum, L.   Camden.
Eryngium maritimum, C. Bauh.   Camden.
Bupleurum protractum, Link.   Camden.
Carum Carui, L.   Communipaw, Camden.
Apium leptophyllum, F. Müll.   Communipaw.
Apium graveolens, L.   Camden.
Apium repens, Reich.   Camden.
Coriandrum sativum, L.   Communipaw, Camden.
Chærophyllum temulum, L.   Communipaw.
Fœniculum vulgare, Gærtn.   Camden.
Ammi visnaga, L.   Camden.
Scandix Pecten-Veneris, L.   Communipaw, I. C. Martindale.
Richardsonia scabra, St. Hill.   Communipaw, Camden.

Galium tricorne, With.   Communipaw, Camden. (?)
Sherardia arvensis, L.   Camden.
Valerienella dentata, Koch.   Camden, I. C. Martindale.
Calycera balsamitæfolia, Rich.   Camden, I. Burk.
Buthalmum salicifolium, L.   Hoboken.
Ageratum conyzoides, L.   Camden.
Eupatorium cannabinum, L.   Camden.
Conyza albida, Less.   Hoboken.
Mikania gymnocladus.   Camden, I. Burk.
Heterotheca scabra, DC.   Camden.
Erigeron acre, L.   Camden.
Baccharis Douglassii, DC.   Camden.
Inula dysenterica, L.   Hoboken, Camden.
Inula pulicaria, L.   Hoboken.
Parthenium hysterophorus, L.   Camden.
Eclipta erecta, L.   Communipaw.
Acanthospermum hispidum, DC.   Communipaw.
Melanthera deltoidea, Michx.   Camden.
Hemizonia pungens, T. & G.   Camden.
Bidens leucantha, Willd.   Communipaw.
Helenium quadridentatum, Sabill.   Camden.
Achillea Ptarmica, L.   Communipaw.
Achillea rosea, Desf.   Camden.
Anthemis nobilis, L.   Camden.
Anthemis tinctoria, L.   Camden.
Chrysanthemum segetum, L.   Camden.
Chrysanthemum coronarium, L.   Camden.
Matricaria Chamomilla, L.   Camden, Communipaw.
Cenia turbinata, Pers.   Communipaw.
Tussilago Farfara, L.   Camden.
Senecio Jacobæus, L.   Communipaw, Camden.
Calendula arvensis, L.   Camden.
Onopordon acanthium, L.   Hoboken, I. C. Martindale ; Camden.
Carduus acanthoides, L.   Hoboken, I. C. Martindale ; Camden.
Cnicus pycnocephalus, Jacq.   Camden.
Centaurea Phrygia, L.   Communipaw.
Centaurea Jacea, L. (?)   Communipaw, J. Schrenck.
Contaurea solstitialis, L.   Camden ; Communipaw, J. Schrenck.
Scolymus Hispanicus, L.   Hoboken, Camden.
Lampsana communis, L.   Communipaw, Camden.
Hypochæris radicata, L.   Communipaw, Camden.
Cichorium Endivia, L.   Communipaw.
Cichorium divaricatum, Schweb.   Camden.

Helmintha echioides, Gærtn. Camden.
Picris hieracioides, L. Communipaw, I. C. Martindale.
Crepis virens, L. Hoboken, Camden.
Hieracium umbellatum, L. Communipaw.
Leontodon hirtum, L. Camden.
Lactuca Scariola, L. Communipaw.
Sonchus tenerrimus, L. Communipaw, Camden.
Specularia Speculum, DC. Communipaw.
Asperugo procumbens, L. Communipaw.
Heliotropium Curassavicum, L. Communipaw, Camden.
Heliotropium anchusæfolium, Poir. Communipaw.
Heliotropium supinum, L. Camden.
Heliotropium Indicum, L. Communipaw, Camden.
Tournefortia heliotropioides, Hook. Camden.
Anchusa officinalis, Don. Camden.
Anchusa leptophylla, R. & S. (?) Communipaw.
Echium violaceum, L. Communipaw.
Myosotis collina, Hoff. Communipaw.
Dichondra repens, Forst. Communipaw.
Ipomœa commutata, R. & S. Communipaw.
Ipomœa hederacea, Jacq.; *Var.* integriuscula, Gray. Camden.
Ipomœa lacunosa, L. Camden.
Convolvulus pentapetaloides, L. (?) Camden.
Solanum miniatum, Bernh. Hoboken.
Solanum nigrum, L.; *Var.* Dillenii, Gray. Communipaw.
Solanum nigrum, L.; *Var.* nodiflorum, Gray. Camden.
Solanum gracile, Link. Camden.
Solanum sisymbriifolium, Lam. Camden.
Physalis Peruviana, Nees. Camden, I. C. Martindale.
Hyoscyamus albus, L. Communipaw, I. C. Martindale.
Nicotiana glauca, Grah. Camden.
Atropa Belladonna, L. Camden, I. C. Martindale.
Petunia parviflora, Juss. Camden.
Linaria striata, DC. Communipaw.
Linaria spuria, Mill. Communipaw, Camden.
Linaria minor, Desf. Camden, Communipaw.
Linaria Cymbalaria, Mill. Camden.
Verbascum virgatum, With. Communipaw.
Scrophularia aquatica, L. Camden.
Scrophularia canina, L. Communipaw, J. Schrenck.
Scoparia flava, C. & S. Camden, Communipaw.
Sesamum Indicum, L. Camden, Communipaw.
Lippia nodiflora, Michx. Camden.

Lippia Bonariensis. Camden.
Salvia verbenacea, L. Communipaw.
Leonurus Siberica, L. Communipaw.
Galeopsis versicolor, Curt. Communipaw, Camden.
Stachys annua, L. Communipaw, Camden, Hoboken.
Stachys recta, L. Communipaw, I. C. Martindale.
Stachys sylvatica, L. Camden.
Stachys hirta, L. Camden.
Teucrium Scordium, L. Communipaw, M. Ruger.
Plantago Coronopus, L. Camden.
Plantago lagopus. Camden, I. C. Martindale.
Amarantus deflexus, L. Camden, Communipaw.
Amarantus Blitum, L. Camden.
Amblogyna polygonoides, Raf. Camden.
Cladothrix lanuginosa, Moq. Communipaw.
Chenopodium Vulvaria, L. Camden ; Communipaw, M. Ruger.
Chenopodium obovatum, Moq. Camden.
Roubieva multifida, L. Camden.
Beta maritima, L. Camden.
Beta procumbens, Ch. Smith. Camden.
Atriplex laciniata, L. Hoboken, Camden.
Atriplex rosea, L. Camden, Communipaw.
Blitum rubrum, Reich. Camden.
Rumex aquaticus, L. Camden.
Rumex pulcher, L. Camden.
Celosia cristata, L. Communipaw.
Parietaria officinalis, L. Communipaw, Camden.
Parietaria diffusa, M. & K. Camden.
Ricinus communis, L. Communipaw, Camden.
Acalypha Poiretii, Spr. Camden.
Mercurialis annua, L. Communipaw, Camden.
Euphorbia Peplis, L. Communipaw.
Euphorbia exigua, L. Camden.
Euphorbia segetalis, L. Camden.
Phyllanthus polygonoides, Spr. Communipaw.
Anthericum ramosum, L. (?) Communipaw.
Cyperus umbellatus, Vahl. Communipaw.
Scirpus mucronatus, L. Camden.
Scirpus setaceus, L. Camden.
Fimbristylis congesta, Torr. Camden.
Alopecurus agrestis, L. Communipaw, Camden.
Sporobolus Indicus, Br. Communipaw, Camden.
Agrostis Spica-venti, L. Camden.

Agrostis verticillata, Vill.   Camden.
Eleusine Indica, Gærtn.; *Var.* brachystachys, Trin.   Communipaw.
Eleusine rigida, Spr.   Camden.
Eleusine coracana, Gærtn.   Camden.
Glyceria procumbens, Curt.   Camden.
Festuca spectabilis, Jan.   Camden.
Briza minor, L.   Camden.
Corynephorus canescens, Beauv.   Hoboken, J. Schrenck.
Lepturus incurvatus, Trin.   Camden.
Hordeum murinum, L.   Communipaw, Camden.
Holcus mollis, L.   Camden.
Avena fatua, L.   Camden.
Hierochloa australis, K. & S.   Communipaw.
Phalaris intermedia, Bosc.   Camden.
Phalaris paradoxa, L.   Camden.
Panicum miliaceum, L.   Camden, Communipaw.
Setaria setosa, Swartz.   Camden.

ALSO :

Equisetum variegatum, Schlecht.   Communipaw.

CRYPTOGAMIA.

—♦—

Class III.—ACROGENOUS CRYPTOGAMS.

FILICES.

Polypodium, L. . . . . Polypody.

P. vulgare, L.   Common Polypod.   Banks of the Delaware River below Gloucester, C. F. Parker; and common on rocks in the northern and middle counties.

Adiantum, L. . . . . Maidenhair Fern.

A. pedatum, L.   Common Maidenhair.   Rich moist woods.   Common in the northern and middle counties, but scarce on the Yellow Drift.   Keyport, Monmouth Co., R. W. Brown.

Pteris, L. . . . . Brake.   Bracken.

P. aquilina, L.   Common Brake.   Thickets and hill-sides.   Common throughout the State.

Var. caudata, Hook.   Tailed Brake.   Pine barren regions, and confined to the Yellow Drift.   Brown's Mills, Burlington Co., and near Camden, C. F. Parker; near Tom's River, Ocean Co., N. L. Britton.

Cheilanthes, Swartz. . . . . Lip–fern.

C. vestita, Swartz.   Lip-fern.   Palisades, C. F. Austin ; abundant near Milford, Hunterdon Co., A. P. Garber ; Warren Co., C. F. Parker; Snake Hill, W. H. Leggett.

Pellæa, Link. . . . . Cliff–brake.

P. gracilis, Hook.   Graceful Cliff-brake.   Rocks in a ravine, Godwinville, C. F. Austin, in herb C. F. Parker.

P. atropurpurea, Link.   Purple Cliff-brake.   Sparingly on rocks in the northern counties.   Limestone cliffs between Newton and Swartswood Lake, Wm. Bower ; Sussex Co., C. F. Austin ; Andover, W. H. Rudkin ; near Blairstown, Warren Co., H. H. Rusby.

Woodwardia, Smith. . . . . Chain–fern.

W. Virginica, Smith.   Virginian Chain-fern.   Frequent in pine barren swamps.   Occurs also at Bergen Neck and Carlstadt, and in Bergen Co., W. H. Leggett.

W. angustifolia, Smith.   Narrow-leaved C.   Franklin, Essex Co.,
H. H. Rusby; Bergen Neck, W. H. Leggett; and frequent in swamps
on the Yellow Drift.

Aspelenium, L. . . .   . SPLEENWORT.

A. ebenoides, R. R. Scott.   Newton, Sussex Co., H. H. Rusby in
Bull. Torr. Bot. Club, vii., p. 29.

A. Trichomanes, L.   Frequent on rocks northern and middle counties.   Rare or absent on the Yellow Drift.

A. ebeneum, Ait.   Banks and rocky woods.   Near Camden, C. F.
Parker; and frequent in the northern counties.

A. montanum, Willd.   Cliffs facing the Delaware River, near the
summit of Mt. Tammany, Warren Co., S. W. Knipe.

A. Ruta-muraria, L.   Sparingly on rocks in the northern counties.
Limestone cliffs between Newton and Swartswood Lake, Wm. Bower;
Sussex Co., C. F. Austin, A. P. Garber; near Blairstown, Warren Co.,
H. H. Rusby.

A. thelypteroides, Michx.   Rich woods.   Frequent in the northern
and middle counties.

A. Filix-fœmina, Bernh.   Moist woods.   Common in the northern
and middle counties, and sparingly on the Yellow Drift.   Camden and
Gloucester Cos., C. F. Parker.

Camptosorus, Link.   . . . .   WALKING–LEAF.

C. rhizophyllus, Link.   Walking-leaf.   Scarce.   Palisades, C. F.
Austin; banks of the Delaware River, Warren Co., C. F. Parker;
First Mt. near Plainfield, F. Tweedy, I. C. Russell; east shore of
Swartswood Lake, Wm. Bower; Newton, Sussex Co., H. H. Rusby;
Preakness, W. L. Fischer; Mine Hill, Sussex Co., Prof. G. H. Cook.

Phegopteris, Fée. . . . . BEECH-FERN.

P. hexagonoptera, Fée.   Woods.   Rather common throughout.

Aspidium, Swartz.   . . SHIELD–FERN.   WOOD–FERN.

A. Thelypteris, Swartz.   Marshes.   Quite common throughout,
especially in the northern parts of the State.

A. Noveboracense, Swartz.   Swamps and moist thickets.   Common
throughout the State.

A. spinulosum, Swartz.; *Var.* intermedium, Eaton.   Woods.   Common throughout the State.

Var. dilatatum, Hook.   Rare, and confined to the northern counties.   Rocky woods, Warren Co., C. F. Parker; Lake Hopatcong,
Morris Co., T. C. Porter.

A. cristatum, Swartz. Closter, C. F. Austin; near Washington, Warren Co., and Camden Co., C. F. Parker; Red Bank, Monmouth Co., A. B. Guilford; Great Swamp, W. H. Leggett.

Var. Clintonianum, Eaton. New Jersey, Gray's Manual.

A. Goldianum, Hook. Damp woods, Hunterdon Co., C. F. Parker; Marble Hill, Warren Co., T. C. Porter; First Mt., Essex Co., H. H. Rusby.

A. marginale, Swartz. Camden Co., C. F. Parker; Monmouth Co., Dr. Torrey in Willis Catalogue; Bergen Point, W. H. Leggett; and common in the northern counties.

A. acrostichoides, Swartz. Woods. Common in the northern and middle counties, and frequent on the Yellow Drift.

Var. incisum, Gray. Bergen Co., G. C. Woolson in Willis Catalogue.

Cystopteris, Bernh. . . . . . . BLADDER-FERN.

C. bulbifera, Bernh. Scarce. Marble Hill, Warren Co., C. F. Parker.

C. fragilis, Bernh. Closter, C. F. Austin; Warren Co., C. F. Parker; common on First Mt., Essex Co., H. H. Rusby; Freehold, O. R. Willis.

Onoclea, L. .   . . . SENSITIVE FERN.

O. sensibilis, L. Common Sensitive Fern. Moist places. Common throughout.

Woodsia, R. Br. . . . . WOODSIA.

W. obtusa, Torr. Palisades, C. F. Austin; Marble Hill, Warren Co., T. C. Porter; common on First Mt., Essex Co., H. H. Rusby; Bergen Point, W. H. Leggett; near Newton, Sussex Co., Arthur Hollick; Little Snake Hill, N. L. Britton.

W. Ilvensis, R. Br. Palisades, C. F. Austin; Franklin, Sussex Co., A. P. Garber; Warren Co., C. F. Parker; First Mt., Essex Co., H. H. Rusby; Marble Hill, Warren Co., T. C. Porter.

Dicksonia, L'Her. . . . . DICKSONIA.

D. pilosuiscula, Willd. (D. punctilobula, Kunze.) Moist woods. Common throughout.

Schizæa, Smith. . . . . SCHIZÆA.

S. pusilla, Pursh. Sparingly in pine barren regions. Quaker Bridge, Dr. John Torrey; Tom's River and near Kettle Creek, Ocean Co., P. D. Knieskern; Pleasant Mills, Atlantic Co., and cedar swamps along the Atsion River, Burlington Co., C. F. Parker; Ferrago, C. F. Austin.

Lygodium, Swartz. . . . . CLIMBING-FERN.

L. palmatum, Swartz. Hartford-fern. Scarce. Near Shark River, Monmouth Co., P. D. Knieskern; near Hightstown, O. R. Willis; Rancocus, W. M. Canby in Willis Catalogue; near Matawan, S. Lockwood; near Keyport and Mount Pleasant, Monmouth Co., R. W. Brown; Brown's Mills, Burlington Co., C. F. Parker; near White Horse, Camden Co., C. E. Smith; Craner's Mills, two miles south of New Brunswick, Prof. Geo. H. Cook.

Osmunda, L. . . . . FLOWERING-FERN.

O. regalis, L. Flowering-fern. Low grounds. Quite common throughout the State.

O. Claytoniana, L. Clayton's Fern. Low grounds. Common in the northern and middle counties.

O. cinnamomea, L. Cinnamon-fern. Swamps and low copses. Common throughout the State.

Var. frondosa, Gray. Occasional. Camden, C. F. Parker.

OPHIOGLOSSACEÆ.

Botrychium, Swartz. . . . . MOONWORT.

B. lanceolatum, Angstr. Borders of swamps in shady places. Closter, Bergen Co., and Chester, Morris Co., C. F. Austin.

B. Virginicum, Swartz. Rich woods. Common in the northern and middle counties. Rare on the Yellow Drift.

B. ternatum, Swartz.; *Var.* obliquum, Milde. Low woods. Quite common throughout.

Var. dissectum, Milde. Frequently found with the last.

Ophioglossum, L. . . . . ADDER'S TONGUE.

O. vulgatum, L. Adder's Tongue. Monmouth Co., Dr. Torrey in Willis Catalogue; Closter, Bergen Co., C. F. Austin; Andover, Sussex Co., A. P. Garber; Budd's Lake, Morris Co., T. C. Porter.

EQUISETACEÆ.

Equisetum, L. . . . . HORSETAIL, RUSH

E. arvense, L. Common Horsetail. Moist places. Common throughout the State.

Var. serotinum, Meyer. Closter, Bergen Co., C. F. Austin.

E. pratense, Ehrh. Near Closter and Sparta, C. F. Austin.

E. sylvaticum, L. Budd's Lake, T. C. Porter; Closter, C. F. Austin; Swartswood Lake, N. L. Britton.

E. limosum, L. Sparingly in the northern counties. Closter, C. F. Austin; near Andover, A. P. Garber; Budd's Lake, T. C. Porter.

E. hyemale, L. Scouring-rush. Rare in Monmouth and Ocean Cos., P. D. Knieskern; Camden, C. F. Parker; and frequent in the northern and middle counties.

LYCOPODIACEÆ.

Lycopodium, L., Spring. . . . . .   CLUB–MOSS.

L. lucidulum, Michx. Near Camden, and banks of Timber Creek, Camden Co., C. F. Parker; and frequent in the northern and middle counties.

L. inundatum, L. Closter, Bergen Co., C. F. Austin.

Var. Bigelovii, Tuckerm. Frequent in low grounds on the Yellow Drift.

L. annotinum, L. Closter, Bergen Co., C. F. Austin.

L. alopecuroides, L. Common in pine barren swamps.

L. dendroideum, Michx. Ground-pine. Moist woods. Rare in Ocean and Monmouth Cos., P. D. Knieskern; Camden, C. F. Parker; and frequent in the northern and middle counties.

L. clavatum, L. Common Club-moss. Woods. Frequent in the northern parts of the State.

L. Carolinianum, L. Frequent in pine barren swamps.

L. complanatum, L. Christmas-green. Woods and thickets. Common throughout the State.

Var. sabinæfolium, Gray. Norwood, Bergen Co., C. F. Austin.

Selaginella, Beauv., Spring. . . .   SELAGINELLA.

S. rupestris, Spring. Rocks, northern and middle counties, but scarce. Closter, C. F. Austin; Milford, Hunterdon Co., A. P. Garber; common in Essex Co., H. H. Rusby.

S. apus, Spring. Low grounds. Quite common throughout.

ISOETEÆ.

Isoetes, L. . . . .   QUILLWORT.

I. echinospora, Durieu; *Var.* Braunii, Engelm. Lake Hopatcong, Morris Co., T. C. Porter; Tom's River, Ocean Co., C. F. Parker.

I. riparia, Engelm. Gravelly shores of the Delaware River at Camden, C. F. Parker.

I. Engelmanni, Braun. Closter, Bergen Co., C. F. Austin.

Class IV.—ANOGENS.

MUSCI.

COMPILED BY MR. C. F. PARKER FROM THE COLLECTIONS OF THE LATE
COE F. AUSTIN.

Sphagnum, Dill.

  S. Portoricense, Hampe. Manchester pond, Ocean Co.

  S. Austini, Sulliv. Swamps near Farrago and Manchester, Ocean Co.

  S. cymbifolium, Ehrh. Bogs.

    Var. pycnocladum, Aust. Pine barrens.

    Var. squarrosulum, Aust. Closter, Bergen Co.

  S. rigidum, Schimp. Low sandy places, pine barrens.

    Var. humile, Schimp. Pine barrens.

  S. molle, Sulliv. South Jersey.

  S. molluscum, Bruch. Peat bogs, in water one inch or less in depth in pine barrens.

  S. acutifolium, Ehrh. Bogs.

    Var. confertum, Aust. Open bogs.

    Var. purpureum, Aust. Peat bogs, common.

    Var. fuscum, Aust. Peat bogs.

    Var. robustum, Aust. Cedar swamps about Farrago, Ocean Co.

  S. fimbriatum, Wils. Swamps on the Palisades near Closter.

  S. Girgensohnii, Russov. Swamps and bogs, northern part of the State.

  S. teres, Angstr. Marshes, Budd's Lake and South Jersey (rare).

  S. Pylaesii, Brid. Border of pond, Manchester, Ocean Co.

  S. cyclophyllum, Sulliv. & Lesqx. Pools, pine barrens.

  S. neglectum, Angstr. In an open grassy bog near Closter.

  S. subsecundum, N. & H. Meadows, Bergen Co.; pine barrens.

    Var. Lescurii, Aust. Borders of streams, pine barrens.

  S. cuspidatum, Ehrh. Inundated bogs &c., Closter and pine barrens.

    Var. laxifolium, Aust. In pools, Closter; pine barrens.

    Var. Torreyanum, Aust. Manchester pond, Ocean Co.

    Var. plumosum, Aust. Manchester pond, Ocean Co.

    Var. parvum, Aust. In about an inch of water with **S.** molluscum, Bruch., in cranberry bogs, pine barrens.

    Var. recurvum, Aust. (major). Cedar swamps, pine barrens.

  S. macrophyllum, Bernh. Manchester pond, Ocean Co.

  S. papillosum, Lindb. Bogs near Batsto, Atlantic Co.

Andræa, Ehrh.

    A. petrophila, Ehrh.   On steep dry rocks and boulders near Closter and Sparta.

    A. rupestris, Turner.   Delaware Water Gap, New Jersey side.

Archidium, Brid.

    A. Ohioense, Schimp.   Flat rocks, Palisades, Bergen Co.

Micromitrium, Aust.   (Namomitrium, Lindbg.)

    M. Austini (Sulliv.), Aust.   Closter, Bergen Co.

    M. synoicum, James.   On surface of clods of clay, Camden; James, Austin.

    M. megalosporum, Aust.   With the preceding, (very rare).

Ephemerum, Hampe.

    E. serratum, Schreb.   Fields and gardens near Closter.

    E. crassinervium, Schwæger.   Damp ground, Closter.

    Var. spinulosum, Aust.   Palisades, Camden.

    E. papillosum, Aust.   Rocks, Palisades.

Sphærangium, Sch.

    S. muticum, (Schreb.,) Schimp.   Rocks, Palisades.

    S. triquetrum, Schimp.   Sandy fields, Tom's River.

Phascum, L.

    P. cuspidatum, Schreb.   Old fields.

Pleuridium, Brid.

    P. alternifolium, Brid.   Old fields, Bergen Co.; Camden.

Sporledera, Hampe.

    S. palustris, Schimp.   Old fields, N. J. (?)

Bruchia, Schwægr.

    B. flexuosa, Schwægr.   Old fields, Bergen Co.

Systegium, Br. Eu.

    S. nitidulum, Schimp.   Old fields.

    S. Sullivanti, Schimp.   Old fields, Closter.

Gymnostomum, Hedw.

    G. rupestre, Schwægr.   Damp shaded rocks, Palisades.

    G. curvirostrum, Hedw.   On wet rocks, Godwinville.

Hymenostomum, R. Br.

　H. microstomum, R. Br.　Rocky ground, Palisades.

Weisia, Hedw.

　W. viridula, Brid.　Old fields, &c., common.

　, **W.** serrulata, Funk.　On the perpendicular face of moist rocks,
　　Palisades and Del. Water Gap.

Rhabdoweisia, Br. & Sch.

　R. denticulata, Brid.　On rocks in ravines, Pascack.

Dicranum, Hedw.

　D. rufescens, Turner.　On naked banks, &c., throughout North
　　Jersey.

　D. pellucidum, Hedw.　On rocks subject to inundation in deep
　　glens, near West Vernon, Sussex Co.

　D. Schreberi, Hedw.　Wet rocks and banks, Hohokus.

　D. varium, Hedw.　Moist banks, Closter.

　D. heteromallum, Hedw.　Moist grounds, common.

　　Var. orthocarpon, Aust.　Moist ground and banks.

　D. montanum, Hedw.　N. J. (?)

　D. flagellare, Hedw.　On logs in woods, Closter.

　D. fulvum, Hook.　N. J. (?)

　D. longifolium, Hedw.　On shaded rocks and trunks of trees,
　　North Jersey.

　D. scoparium, Hedw.　On the ground in woods, Closter.

　　Var. orthophyllum, Schimp.　Palisades, Bergen Co.

　　Var. curvulum, Schimp.　Palisades, Bergen Co.

　　Var. orthocarpum, Aust.　Woods, near Closter.

　　Var. minor, Aust.　Sandy banks, near Closter.

　　Var. rupestre, Suliv. & Lesqx.　On granite and trap rocks, in
　　　the mountains of N. J.

　　Var. paludosum, Schimp.　Springy places in swamps, near
　　　Closter.

　D. Schraderi, Schw.　Swamps and wet woods, common.

　D. spurium, Hedw.　Rocks on Mts., North Jersey.

　　Var. (**D** condensatum, Hedw.)　Near Coleville, and in white
　　　sand, pine barrens; common.

　D. undulatum, Turner.　On rocks and on the ground, Palisades;
　　near Baumpie's Hook.

　D. robustum, Blytt.　Swamps about Closter (sterile); and in pine
　　barrens, James.

Trematodon. Rich.

    T. longicollis, Michx. Low grounds, Closter (rare).

Leucobryum, Hampe.

    L. glaucum, Linn. On the ground in woods, common.

Fissidens, Hedw.

    F. bryoides, Hedw. On the ground in woods, Closter.
    F. incurvus, Schw.; *Var.* (**F.** minutulus, Sulliv.) On stones in damp shady or springy places, Closter.
    F. osmundioides, Hedw. Springy places in swamps, Closter.
    F. subbasilaris, Hedw. On roots of trees in woods, Closter; on limestone rocks, Sussex Co.
    F. adiantoides, Hedw. On wet rocks, banks, &c., very common.
    F. taxifolius, Hedw. On the ground in woods, Closter.
    F. Closteri, Aust. On rocks along rivulets, Palisades near Closter, Bergen Co.

Conomitrium, Mont.

    C. Julianum, Mont. Rocky streams, North Jersey.
    C. Hallianum, Sulliv. & Lesxq. On shaded rocks moistened by spray, at Little Falls and Ogdensburg.

Campylostelium.

    C. saxicola, Web. & Mohr. On sandstone boulders, near Closter, rare.

Seligeria, Br. & Sch.

    S. recurvata, Hedw. On moist shaded rocks at Hohokus and Godwinville.

Pottia, Ehrh.

    P. riparia, Aust. On moist rocks along streams, Palisades and Northern New Jersey.

Didymodon, Br. & Sch.

    D. rubellus, Roth. On rocks along streams, Northern New Jersey.
    D. cylindricus, Bruch. Moist rocks and banks, Northern New Jersey (rare).

Ceratodon, Brid.

    C. purpureus, Brid. On the ground; very common.
    Var. aristatus, Aust. In sandy pine barrens.

Trichostomum, Br. & Sch.

    T. tortile, Schrad. Roadsides. Frequent. Closter.

    T. lineare, Swartz. Roadsides, &c., Camden, C. F. Parker; Closter, Austin.

    T. pallidum, Hedw. On the ground. Very common.

    T. glaucescens, Hedw. Crevices of rocks, Little Falls.

Desmatodon, Brid.

    D. arenaceus, Sulliv. & Lesqx. On the ground, Closter and South Jersey.

Barbula, Hedw.

    B. unguiculata, Hedw. About the roots of trees, Closter.

    B. fallax, Hedw. Closter and North Jersey, (rare).

    B. cæspitosa, Schwægr. About the roots of trees. Common.

    B. tortuosa, Web. & Mohr. New Jersey. (?)

    B. fragilis, Hook. & Wils. On dry limestone rocks at the New Jersey Zinc Mines. Sterile.

    B. muralis. Hedw. On old walls, Palisades and central part of the State.

    B. ruralis, Hedw.; *Var.* rupestris, Schimp. On rocks at the base of the Palisades; also about the Zinc Mines, Sussex Co.

    B. papillosa, Wils. Trunks of Buttonwood trees, Batsto, James ; limestone rocks about the Zinc Mines, Sussex Co.

Tortula, Hedw.

    T. recurvifolia, Schimp. On rocks, Hoboken.

Grimmia, Ehrh.

    G. apocarpa, Hedw. On rocks in moist ravines. Common.
        Var. gracilis, Aust. On shaded rocks, Palisades and Greenwood Mts.

    G. conferta, Funk. On rocks and on the ground, Passaic Falls.
        Var. I. On rocks and on the ground, limestone region of New Jersey.
        Var. II. Dry limestone rocks near Sparta.

    G. Pennsylvanica, Sch. On rocks, Closter, Bergen Co.

    G Olneyi, Sulliv. On exposed rocks, Palisades and northern New Jersey.

    G. leucophæa, Grev. Exposed rocks, Palisades. Sterile.

Racomitrium, Br. & Sch.

    R. aciculare, Brid. Rocky beds of streams, Palisades.

R. Sudeticum, Brid. On irrigated rocks, North Jersey.

R. microcarpum, Brid. Exposed rocks, Mts. North Jersey (rare).

Hedwigia, Ehrh.

H. ciliata, Ehrh. Exposed rocks and boulders, common.

Var. Inundated rocks in the Passaic River near Paterson,
and Little Falls.

Drummondia, Hook.

D. clavellata, Hook. On the trunks and branches of trees (particularly Juniperus Virginiana), &c.; very common.

Ptychomitrium, Br. & Sch.

P. incurvum, Sch. On old stone fences (rarely on rocks) Closter.

Amphoridium, Sch.

A. Lapponicum, Hedw. Crevices of rocks, North Jersey.

Orthotrichum, Hedw.

O. Ludwigii, Sch. On trees and stone fences, Palisades.

O. Hutchinsiæ, Hook & Tayl. On dry rocks, Closter.

O. crispum, Hedw. On trees, Closter.

O. crispulum, Hornsch. On trees, particularly Betula lutea, on the Palisades.

O. anomalum, Hedw. Rocks, Palisades.

O. Peckii, Sulliv. & Lesqx. On dry limestone rocks, Sussex Co.

O. Lescurii, Aust. On dry shaded rocks, Northern Jersey.

O. pumilum, Swartz. On shade trees in the towns of Central N. J.

O. strangulatum, Beauv. On trees, Closter.

O. sordidum, Sulliv. & Lesqx. On trees, common.

O. citrinum, Sulliv. & Lesqx. On trees, N. J.

O. leiocarpum, Bryol. Europ. On Juniperus Virginiana, Palisades.

O. psylocarpum, James. Central N. J. on shade trees.

Tetraphis, Hedw.

T. pellucida, Hedw. On rotten wood; common.

Encalypta, Schrad.

E. streptocarpa, Hedw. On limestone rocks, Sussex Co.

Tetraplodon, Br. & Sch.

T. australis, Sulliv. & Lesqx. Pine barrens and Tom's River.

Aphanorrhegma, Sulliv.

A. serrata, Hook. & Wils. On damp ground, common.

Physcomitrium, Brid.

P. immersum, Sulliv.   Low banks of the Delaware, Camden.

P. pyriforme, Brid.   On damp ground, Closter.

Funaria, Schreb.

F. hygrometrica, Hedw.   On the ground, Closter.

Var. patula, Austin.   Chiefly on damp walls, N. J.

F. flavicans, Michx.   In pastures, on small patches of ground where it is killed by urine, N. J.

Bryum, Br. & Sch.

B. pyriforme, (Linn.,) Hedw.   Pine barrens.

B. nutans, Schreb.; *Var.*   On exposed rocks, mountain-tops.

B. crudum, Schreb.   Banks and ravines, sterile.

B. albicans, Wahlenb. (B. Wahlenbergii, Sch.)   On wet rocks and banks, Pascack.

B. Lescurianum, Sulliv.   On banks along roadsides, &c., Palisades.

B. bimum, Schreb.   Wet places, common.

Var. ———.   Wet rocks, Little Falls, Passaic Co.

B. cernuum, Hedw.   About the roots of trees in open woods, near Closter.

B. cæspiticium, Linn.   On the ground, &c., very common.

B. argenteum, Linn.   On the ground, old roofs, &c., common.

B. capillare, Linn.   Shaded banks, rocks, &c., common; sterile in N. J.

B. pseudo-triquetrum, Sch.   Moist rocks, very common in North Jersey.

Var. ———.   Wet rocks, Palisades (rare).

B. roseum, Schreb.   On old logs, about the roots of trees, &c., Closter.

Mnium, Br. & Sch.

M. cuspidatum, Hedw.   On damp shaded ground, rocks, &c., common.

M. affine, Bland.   Swamps and wet rocks, common.

M. medium, (?) Byol. Europ.   Wet rocks, Little Falls.

M. rostratum, Sch.   Wet rocks, Palisades and Northern; rare.

M. lycopodioides, Hook.   On shaded rocks, Northern N. J.

M. spinulosum, Bryol. Europ.   On rocks in a ravine at Godwinville.

M. serratum, Brid.   Shaded banks and crevices of rocks, Palisades and Northern N. J.

M. hornum, Hedw.   Cedar swamps, near New Durham, and rocky or sandy banks of streams.

M. punctatum, Hedw.   Swamps, common.
   Var. ———.   Banks of rivulets, common.
M. cinclidioides, Blytt.   Swamps and wet woods, Palisades (fre-
   quent) and Northern N. J.
M. stellare, Hedw.   On shady rocky cliffs and banks, and about
   the roots of trees in swamps, rather frequent.

Meesea, Hedw.

   M. tristicha, Funk.   Wet meadows about Closter.

Aulacomnion, Schwaegr.

   A. palustre (L), Sch.   Swamps and low grounds, common.
      Var. rupestre, Aust.   Moist rocks, Palisades, near Closter.
   A. heterostichum, Bryol. Europ.   On banks and about the roots
      of trees in woods, common.

Bartramia, Hedw.

   B. fontana, Brid.   Banks of rivulets and in springy places,
      common.
   B. pomiformis, Hedw.   Same situations as the preceding, very
      common.
   B. Œderi, Swartz.   Banks of ravines, northern N. J.

Timmia, Hedw.

   T. megapolitana, Hedw.   Banks of ravines, northern N. J.

Atrichum, Beauv.

   A. angustatum, Brid.   On the ground in open woods and on
      banks, common.
   A. undulatum, Beauv.   Shaded banks, common.
   A. crispum, James.   Banks of rivulets, near Camden, T. P. James,
      C. F. Parker; Tom's River, Ocean Co., Austin.

Pogonatum, Beauv.

   P. brevicaule, Brid.   On banks, roadsides, &c.   Common.
   P. brachyphyllum, Beauv.   On sandy loam along roadsides near
      Tom's River, Austin; Woodbury, Gloucester Co., James.

Polytrichum, Brid.

   P. commune, Linn.   On the ground in woods and old sterile
      fields.   Very common.
   P. formosum, Linn.   About the roots of trees in damp woods
      and on flat shaded rocks.   Common.
   P. juniperinum, Hedw.   On the ground in dry exposed places.
      Common at Closter and in pine barrens.

P. strictum, Menzies.   Sandy pine barrens.   Common.

P. piliferum, Schreb.   In old fields (fertile) and on rocks (sterile) near Closter.

Diphyscium, Web. & Mohr.

D. foliosum, Web. & Mohr.   On banks in woods and rocky ravines.   Common.

Buxbaumia. Haller.

B. aphylla, Haller.   On the ground in open woods.   Rare.   Closter and pine barrens.

Fontinalis, Dill.

F. antipyretica, L.; *Var.* gigantea, Sulliv.   In rivulets.   Common.   Sterile.

F. Novæ-Angliæ, Sulliv.   In springs and in rivulets in swampy places; common.   Sterile.

F. Lescurii, Sulliv.   Ponds and sluggish streams, southern New Jersey.   Sterile.

Var. ——.   In rocky rivulets.   Common.

Var. (?) cymbifolia, Aust.   Ponds of northern N. J.   Sterile.

F. Sullivanti, Lindb.   (F. Lescurii; *Var.* gracilescens, Sulliv.)   Stagnant pools in woods about Closter.

F. Dalecarlica, Bryol., Europ.   Rocky rivulets.   Common.

F. disticha, H. & W. (?)   Rocky rivulets, Mts. of N. J.

Dichelyma, Myrin.

D. capillaceum, Dill.   Swamps, Closter.

Cryphæa, Mohr.

C. glomerata, Schimp.   On Red Cedars, Palisades, rare.

Leptodon, Mohr.

L. trichomitrion, Mohr.   On trees and rocks, common.

Neckera, Hedw.

N. pennata, Hedw.   On trees and rocks, common.

Homalia, Brid.

H. gracilis, James.   Under overhanging rocks, Palisades.

Leucodon, Schwægr.

L. julaceus, (Hedw.,) Sch.   On trees, common.

Thelia, Sulliv.

 T. hirtella, Hedw. On the trunks of trees near the ground, common ; rarely on rocks.

 T. asprella, Schimp. On the roots of trees, old stumps, and on stones in open woods, common.

 T. Lescurii, Sulliv. On flat rocks, Palisades ; on white sand about the base of stunted oaks, in southern N. J.

Myurella, Bryol., Europ.

 M. Careyana, Sulliv. Banks of ravines about Hohokus and in the mountains of New Jersey ; rare in fruit.

Leskea, Hedw.

 L. denticulata, Sulliv. On dry rocks and roots of trees, sterile.

 L. Austini, Sulliv. On stone fences, Sussex Co.

 L. obscura, Hedw. On the roots of trees, and on stones within the reach of floods in low grounds, common.

Anomodon, Hook. & Tayl.

 A. rostratus, (Hedw.,) Schimp. (Leskea rostrata, Hedw.) About the roots of trees in woods, forming dense cushions, Closter.

 A. tristis (Cesati), Hook. & Tayl. On trees and rocks, common ; always sterile.

 A. attenuatus, Hartm. About the roots of trees and on rocks, common.

 A. obtusifolius, Bryol., Europ. On trunks of trees and on rocks, common.

 Var. fragilis, Aust. On trees about Closter.

 A. viticulosus, Linn. On limestone rocks, Sussex Co.

Pterigynandrum, Hedw.

 P. filiforme, Hedw. On rocks and roots of trees, Palisades.

Platygyrium, Bryol., Europ.

 P. repens, Bryol., Europ. On the roots of the Chestnut and Beech, but more commonly on dead wood. Frequent.

Cylindrothecium, Bryol., Europ.

 C. sedutrix, Hedw. On roots of trees, stones, old logs, &c.; also in moist or wet grounds. Very common.

 C. cladorrhizans, Hedw. On old logs, roots of trees, &c., Palisades and northern N. J.

 C. brevisetum, Wils. On leaning trunks of trees, old logs and stone fences ; also on rocks, Palisades and northern N. J.

Climacium, Web. & Mohr.

> C. dendroides, (Dill.) Web. & Mohr. In swamps about Closter
> and Tom's River.
>
> C. Americanum, Brid. On the ground about the roots of trees
> in swamps, and on moist rocks. Very common.
>
> *Var.* fluitans, Aust. In stagnant pools in woods near Closter.

Fabronia, Raddi.

> F. octoblepharis, (Sleich.) Bryol., Europ. On rocks and trees at
> the Delaware Water Gap.

Pylaisia, Bryol., Europ.

> P. subdenticulata, Schimp. On the bases of White Oaks about
> Closter.
>
> P. intricata, (Hedw.) Schimp. On trees. Very common.
>
> *Var.* —— –. Limestone region, northern N. J.
>
> P. velutina, Bryol., Europ. On trees (chiefly young Elms) in
> swamps; also on old logs, &c., in mountainous regions, N. J.
>
> *Var.* ——. On Red Cedars.

Homalothecium, Bryol., Europ.

> H. subcapillatum, (Hedw.,) Bryol. Europ. On trees and old rocks,
> common.

Thuidium, Bryol., Europ.

> T. pygmæum, Bryol., Europ. Mem. On stones along rivulets,
> North Jersey.
>
> T. minutulum, Hedw. On decayed wood in swamps, common.
>
> T. gracile; *Var.* Lancastriense, Sulliv. On dry sterile ground, in
> open woods, common in N. J.
>
> T. scitum, Beauv. On the base of a tree near Closter.
>
> *Var.* aestivalis, Austin. On the roots of trees, N. J.
>
> T. tamariscinum, Hedw. On the roots of trees, old logs, &c., very
> common.
>
> T. delicatulum, Linn. On shaded rocks and banks, common.
>
> T. abietinum, Linn. On dry limestone ridges, Sussex Co., very
> abundant.

Elodium, Sulliv.

> E. paludosum, Sulliv. Swamps and low grounds, common.

Camptothecium, Sch.

> C. nitens, Schreb. In peat-bogs, near Sparta.

Brachythecium, Bryol., Europ.

B. lætum, Brid. On the ground, old logs, roots of trees and rocks, common.

B. acuminatum, Beauv. On the roots of trees and old logs in woods, Closter.

B. salebrosum, Hoffm. On the ground, old wood, &c., common.

B. campestre, Bruch. On the ground in woods, Closter and Northern N. J.

Var. ———. On the ground under shrubbery, in yards and gardens, Closter.

B. acutum, Mitten. On the ground in swamps, near Closter.

B. rutabulum, Linn. On wet shaded ground, dripping rocks and old wells, common.

B. rivulare, Bruch. On rocks in rivulets, Palisades; common.

B. Starkii, Brid. On old logs in mountains, Del. Water Gap.

B. plumosum, Linn. On rocks in rivulets and ravines, very common.

B. Novæ-Angliæ, Sulliv. & Lesqx.; Var. rupestre, Austin. On irrigated rocks in mountains of New Jersey.

Eurhynchium, Bryol. Europe.

E. Boscii, Sch. On shaded banks, common.

E. strigosum, Hoffm. On banks in woods, common.

E. diversifolium, Schimp. On shaded banks, N. J. (?)

E. Sullivantii, R. Spruce. On banks of deep shaded ravines, common.

E. hians, Hedw. In low swamps near Closter.

E. piliferum, Schreb. On the ground about the roots of trees and old logs, in swampy places, N. J., rare.

Thamnium, Bryol. Europ.

T. Alleghaniense, C. Mull. In deep crevices of wet rocks (sterile) Palisades.

Rhynchostegium, Bryol. Europ.

R. demissum, Wils. On damp shaded rocks, Palisades (very rare) and northern N. J.

R. microcarpum, C. Mull.; Var. anisocarpum, Sulliv. On stones in damp woods, about Closter, frequent.

R. recurvans, Sch. On decayed wood, &c., Closter.

Var. ———. Cedar swamps, northern N. J.

R. deplanatum, Schimp. On the ground, under rocks in moist ravines, Palisades.

R. geophilum, Aust. (Hypnum depressum, James). On clayey, shaded ground, N. J.

R. serrulatum, Hedw. On the ground, roots of trees, &c., in woods and swamps, common.

R. rusciforme, Weis. On rocks in rapid streams, common.

Plagiothecium, Bryol. Europ.

P. elegans, Hook. Crevices of shaded rocks, northern N. J.

Var. terrestre, Lindby. On the ground, in a ravine near Pascack (sterile).

P. Mullerianum, Schimp. Rocky ravines, N. J.

P. Passaicense, Aust. Rocky banks, Passaic, Morris and Bergen Counties.

P. latebricola, Wils. About the roots of old stumps, &c., in swamps near Closter.

P. turfaceum, Lindbg. On the ground in woods, &c., Palisades.

P. striatellum, Brid. (**P.** Muhlenbeckii, Bryol., Europ.) Crevices of rocks and rocky banks, Closter.

Var. chrysophylloides, Schimp. On the ground in woods and swamps. Common in N. J. Closter, Austin; Camden Co., Parker.

P. denticulatum, Dill.; Var. ———. On the ground in wet woods and swamps. Common.

Var pusillum, Aust. On flat rocks in the shade of Hemlocks; also on the roots of trees in dry woods, N. J.

P. sylvaticum, Linn. Deep wooded ravines. Mountains of N. J.

Var. 1. (**P.** Sullivantiæ, Schimp.) Crevices of rocks and rocky banks. Common. N. J. (?)

Var. 2. (**P.** Roseanum, Hampe.) On tussocks and about the roots of trees in swamps. Common in N. J.

P. (?) subfalcatum, Aust. Crevices of rocks, Mts. of N. J.

Amblystegium,

A. confervoides, Brid. On limestone rocks, N. J.

A. adnatum, Hedw. On stones and roots of trees. Very common.

A. Lescurii, Sulliv. On rocks in mountain rivulets, N. J., (rare).

A. serpens, Var. irriguum, (Hook. & Wils.) Aust. On wet rocks, &c., at Little Falls.

Var. radicale subjulaceum. On limestone fences along dusty highways. Sussex Co.

Var. radicale parvulum. On trunks of trees in dry woods.

Var. orthocladon major. Springy places in swamps. Closter.

Var. orthocladon fontanum. In limestone springs.

A. fluviatile, Swartz. On rocks in a rivulet near Closter. Sterile.

A. riparium, L.; *Var.* ———. In rivulets and springs. Common.

Var. ———. Inundated places in swamps, Closter.

Hypnum, Dill.

H. Bergenense, Aust. On decaying leaves, &c., about Closter. Very common.

H. hispidulum, Brid. On the ground, roots of trees, dead wood, rocks, &c., common.

H. chrysophyllum, Brid. (H. polymorphum, Hook & Tayl.) On the ground in fields and woods, very common.

 Var. rupestre, Aust. On shaded or dripping rocks, common, N. J. (?)

 Var. uncinifolium, Aust. Moist rocky banks, Hohokus.

H. stellatum, Schreb. Wet meadows, near Closter.

 Var. protensum, (Brid.) Aust. Bogs and swamps, near Closter.

H. polyganum, Schimp. (Amblystegium polyganum, Bryol. Europ.) Swamps about Closter, rare.

H. aduncum, *Var.* 1. On wet rocks, Little Falls.

 Var. 2. Marshy places, Sussex Co.

 Var. gracilescens, Bryol. Europ. On the ground in exsiccated places, common, N. J. (?)

 Var. giganteum, Bryol. Europ. In Budd's Lake, Morris Co., Austin, T. C. Porter.

H. Kneiffii, Bryol. Europ. In sunken places, about Closter.

H. uncinatum, Hedw. On an old stone fence near Closter.

H. pallescens, Schimp. On Kalmia latifolia, in swamps, mountains of N. J.

H. reptile, Michx. On the roots of trees, decayed logs, and on stones near the ground, very common.

 Var. viride, Aust. On roots of trees in woods, Closter.

H. imponens, Hedw. On decayed logs in woods and among Sphagna in swamps; also on white sand in the dry pine barrens; very common.

H. cupressiforme, Linn.; *Vars.* ———. On rocks, roots of trees, &c.

H. curvifolium, Hedw. On decayed woods, wet ground and on rocks, very common.

H. pratense, Koch On the ground in swampy places, common.

 Var. (?) On tussocks and old logs in cedar swamps, near New Durham.

H. micans, Swartz. N. J. (?)

 Var. pulvum, (H. K. & Wils.,) Aust. On inundated logs, &c., in cedar swamps, Tom's River, Ocean Co.

Var. albulum, (C. Mull,) Aust. On decaying leaves, &c., margins of stagnant pools, near Closter.

H. Haldanianum, Grev. On the ground, old logs, &c., in woods, common.

H. molluscum, Hedw. On the ground in damp woods. Common. Usually sterile.

H. Crista-castrensis, Linn. On the ground and on decaying logs in deep damp woods and swamps. Closter. Rare.

H. palustre, Linn. On rocks in mountain rivulets. Rare.

H. molle, Dicks. On rocks along streams, northern N. J. Rare.

H. Closteri, Aust. On rocks along rivulets about Closter.

H. Novæ-Cæsareæ, Aust. On rocks in a small rivulet which crosses the "State Line."

H. cordifolium, Hedw. Swamps. Common.

H. cuspidatum, Linn. Bogs. Common.

H. Schreberi, Willd. On the ground in woods. Very common.

Hylocomium, Bryol., Europ.

H. splendens, Hedw. On the ground in woods, northern N. J. (Sterile.)

H. brevirostre, Ehrh. Deep shaded ravines and swamps. Common. Sterile in N. J.

H. triquetrum, Linn. On the ground in woods. Common. Sterile in N. J.

Rhytidium, Sulliv.

R. rugosum, (Ehrh.) Sulliv. On flat rocks, Palisades. Common. Sterile.

HEPATICÆ.

COMPILED BY MR. C. F. PARKER FROM THE COLLECTIONS OF THE LATE
C. F. AUSTIN.

Sarcoscyphus, Cord.

S. sphacelatus, Gieske. Wet rocks, mountains of N. J.

Plagiochila, Nees & Montg.

P. purelloides, Torr. Among mosses in swamps and shaded ravines, common.

P. asplenoides, Linn. In rocky shaded rivulets, common. Closter.

Scapania, Lindenberg.

S. compacta; *Var.* irrigua, (Nees,) Aust. Near Tom's River.

S. nemorosa, (Linn.,) Nees. Margins of rivulets, swamps, &c., common.

S. nemorosa, Sulliv. In shady places, on rocks and on the ground, very common.

S. albicans; *Var.* taxifolia, *minor.* On banks in woods; also on rocks and on the ground in damp shady ravines, common.

Leptoscyphus.

L. Taylori; *Var.* Among Sphagna in a peat bog, near Closter.

Southbya, Austin.

S. biformis, Aust. On steep wet rocks, Delaware Water Gap, N. J.

Jungermania, Linn.

J. Schraderi, Mart. On the ground, rotten wood, &c., very common.

J. hyalina, Lyell. On banks in woods, near Closter.

J. sphærocarpa; *Var.* (?) On the banks of a small creek subject to inundation, in low grounds (shaded), near Closter.

J. crenulata, Smith. On the ground in old fields, along road-sides, &c., common.

J. (Solenostomum) crenuliformis, Aust. On rocks in rivulets, near Closter.

J. fossombronioides, Aust. On rocks in a rivulet, near Closter.

J. pumila, With. On shaded rocks along rivulets, about Closter, common.

J. inflata, Huds. Sandy pine barrens, near Batsto, T. P. James.
Var. fluitans, Synop. Hepat. In a peat bog, near Closter.

J. excisa, Dicks. On sterile ground in open woods; common.
Var. crispata, Hook. Shaded banks, on the ground and in crevices of rocks along the Passaic and Delaware Rivers.

J. polita, Nees. In a peat bog, near Closter.

Cephalozia, Dumort.

C. Sullivanti, Aust. On rotten wood, rare.

C. divaricata, *Var.* Pine barrens.
Var. confervoides, Aust. Among Sphagna in a peat bog, near Closter.

C. catenulata, Huben. On rotten wood in swamps, &c.

C. connivens, Dicks. On decaying moss, rotten wood, and on the ground, common and variable.

C. bicuspidata, *Var.* conferta, Aust. On banks in woods, near Closter.

C. curvifolia, Dicks.   On rotten logs in damp woods and swamps;
   common.

Odontoschisma, Dumort.

   O. Sphagni, Dicks.   Among mosses, &c., Closter; old log, Quaker
   Bridge, Thos. P. James.

Lophocolea, Nees.

   L. heterophylla, (Linn.) Nees.   On the ground, old logs, &c., wet
   woods, Closter.

Chiloscyphus, Corda.

   C. polyanthos, (Linn.) Corda.   On the ground, &c., in springy
   places in woods; also on rotten logs in swamps, common;
   Batsto, Thos. P. James.

Calypogeia, Raddi.

   C. Trichomanis, (Dicks.) Corda;   *Var.* rivularis, Aust.   In sluggish streams, or growing in loose turfs on their banks, in
   cedar swamps, Southern, N. J.
   Var. tenuis, Aust.   In a peat bog, near Closter.
   C. Sullivanti, Aust.   On slides at the Delaware Water Gap, Jersey
   side.

Lepidozia, Nees.

   L. reptans, (Linn.) Nees.   On the ground in deep shaded ravines.
   L. setacea, (Web.) Mitt.   On the ground and on rotten wood,
   common; pine barrens.

Mastigobryum, Nees.

   M. trilobatum, (Linn.) Nees.   In deep ravines, wet woods, and
   swamps.   Common.
   Var. 1.   (**M.** tridenticulatum, Mx.)   In swamps.   Common.
   Var. 2.   (Jungermania trilobata, *Var.* Hook.)   On rocks in
   deep ravines.

Trichocolea, Nees.

   T. tomentella, Nees.   Among mosses in swamps and along the
   margin of woodland rivulets.   Common.

Sendtnera, Endl.

   S. juniperina, Swartz.   On rocks, Greenwood Mts.

Blepharozia, Dumort.

   B. ciliaris, (Linn.) Dumort.   On the roots of trees, old logs, &c.,
   old stumps, &c.   Closter.

Blepharostoma, Dumort.

    B. trichophyllum, (Linn.) Dum. On the ground and on rotten wood. Common.

Radula, Nees.

    R. complanata, (Linn.) Dum. On rocks and roots of trees. Common.

    R. obconica, Sulliv. On rocks in ravines. Rare. Bergen Co.

Madotheca, Dumort.

    M. platyphylla, (Linn.) Dumort. On rocks and trees. Closter.

    M. porella, (Dicks.) Nees. On rocks and roots of trees subject to inundation. Common.

Phragmicoma, Dumort.

    P. clypeata, (Schw.) Sulliv. On rocks. Common.

Fruilania, Raddi.

    F. squarrosa, Nees. On rocks, bark of trees, &c.

    F. plana, Sulliv. On shaded rocks.

    F. saxicola, Aust. On inclined surface of dry trap rocks, slightly shaded, near Closter and Little Falls.

    F. Eboracensis, Gottsche. Cedar trees, Palisades. Bergen Co.

    F. Hutchinsiae, (Hook.) Nees. On wet rocks, chiefly in mountain rivulets, Closter.

    F. Grayana, Mont. On rocks and trees, Closter.

Steetzia, Lehm.

    S. Lyelli, (Hook.) Lehm. Among mosses in swamps, often aquatic, common; Farrago Pond, Ocean Co.; Camden.

Pellia, Raddi.

    P. epiphylla, (Linn.) Nees. On the ground along small streams, Closter.

Aneura, Dumort.

    A. pinguis, (Linn.) Dumort. On wet banks, rare.

    Var. ———. In water among Sphagna, South Jersey.

    A. sessilis, Spreng. Old logs partly inundated, in swamps, Closter.

    A. palmata, (Hedw.) Nees. On rotten wood, common.

    A. pinnatifida, Nees. On dripping rocks, Hohokus.

    A. multifida, (Linn.) Dumort. On decaying moss in cedar swamps; common in South Jersey; Closter.

Metzgeria, Raddi.

> **M.** furcata, (Linn.) Nees.   On rocks and roots of trees, very common.

Fossombronia, Raddi.

> **F.** angulosa, Raddi.   Brackish meadows, common.   (Matures in early spring.)
>
> **F.** pusilla, (Linn.,) Nees.   Damp ground, Closter.   (Matures in Sept. and Oct.)
>
> **F.** Cristula, Aust.   On moist sand in unfrequented paths, near Batsto, Atlantic Co.   (Matures in autumn.)

Anthoceros, Mich.

> **A.** lævis, Linn.   On mud in cow tracks, also in cultivated fields, Closter.

Notothylas, Sulliv.

> **N.** valvata, Sulliv.   On wet ground, banks of ditches, &c., common.
>
> **N.** melanospora, Sulliv.   On damp ground, chiefly in cultivated fields, rather common.

Marchantia, Linn.

> **M.** polymorpha, Linn.   Ditches and wet springy places, Closter.

Preissia, Nees.

> **P.** commutata, (Lindbg.,) Nees.   Rocky river bank, North Jersey.

Conocephalus (Fegatella).

> **C.** conicus, (Linn.,) Dumort.   Shady banks of rivulets, common.

Asterella, Pallis.

> **A.** hemispherica, Linn.   Rocky banks, chiefly along streams, common.

Grimaldia, Raddi.

> **G.** barbifrons, Bisch.   Rocky places, near Closter.

Fimbriaria, Nees.

> **F.** tenella, Nees.   On damp ground in old fields, &c., very common.

Riccia, Mich.

> **R.** sorocarpa, Bisch.   Rocky places in unfrequented paths, &c., near Closter.

R. lamellosa, Raddi. Rocky places, Palisades, Bergen Co.

R. arvensis, Aust. Wet broken ground in cultivated fields, &c.,
about Closter.

    Var. hirta, (**R.** hirta, Aust. MS. 1864.) Rocky places, near
Closter.

R. Lescuriana, Aust. Rocky ground in paths, &c., Palisades,
Bergen Co.

R. lutescens, Schw. On broken ground in wet places, &c.,
common.

R. Sullivanti, Aust. On damp or wet broken ground in cultivated fields, Closter.

R. fluitans, Linn. In both stagnant and running water, common;
always sterile.

    Var. terrestris, Aust. On the ground in cultivated fields,
Closter.

R. tenuis, Aust. Wet broken ground, margin of woods, near
Closter.

Class V.—THALLOGENS.

LICHENS.

COLLECTED BY COE F. AUSTIN; NAMED BY PROF. EDWARD TUCKERMAN.

PRINTED FROM A LIST COMPILED BY THE LATE C. F. AUSTIN, IN 1878.

Where no locality is given the plants were collected at Closter,
Bergen County.

Ramalina, Ach., D. N.

    R. calicaris, Fries.

       Var. fastigiata, Fries.

       Var. fraxinea, Fries.

       Var. canaliculata, Fries.

       Var. farinacea, Fries.

    R. rigida, Pers., Tuckm.

Cetraria, Ach., Fries.

    C. Fahlunensis, (L.,) Schær.

    C. Fendleri, Tuck. Ocean Co.

    C. Islandica, Ach. Sussex Co.

C. ciliaris, Ach.
C. lacunosa, Ach.
C. aleurites, Ach.
   Var. placorodia, Tuck.   Ocean Co.
C. juniperina, Tuck.
   Var. virescens, Tuck.
C. aurescens, Tuck.

Evernia, Ach., Mann.
   E. furfuracea, Mann.

Usnea, (Dill.,) Ach.
   U. barbata, Fries.
      Var. florida, Fries.
      Var. strigosa, Ach.
      Var. hirta, Fries.
      Var. rubiginea, Mx.
      Var. plicata, Ach.
      Var. dasypoga, Fries.
      Var. ceratina, Schaer.
   U. trichodea, Ach.
   U. angulata, Ach.

Alectoria, Ach., Nyl.
   A. jubata, Fries.
      Var. chalybeiformis, Ach.
      Var. implexa, Fries.

Theloschistes, (Norm.,) Tuckm.
   T. parietinus, (L.,) Norm.
      Var. lychneus, Nyl.
      Var. polycarpus, Fries.
   T. chrysopthalmus, (L.,) Norm.
   T. concolor, Dicks.

Parmelia, Ach., D. N.
   P. crinita, Ach.
   P. perforata, Ach.
      Var. cetrata, Fries.
   P. perlata, Ach.
   P. tiliacea, Flœrk.
   P. Borreri, Ach.
      Var. rudecta, Tuck.
   P. saxatilis, Ach.
   P. lævigata, Ach.   Ocean Co.

P. pertusa, Schrær.
P. physodes, Ach.
P. colpodes, Ach.
P. caperata, Ach.
P. conspirsa, Ach.
P. ambigua, Ach.
P. olivacea, Ach.

Physcia, (D. C., Fries.,) Th. Fr.

P. aquila, (Ach.,) Nyl.
    Var. detonsa, Tuck.
P. speciosa, (Wulf.,) Fries.
    Var. hypolenca, Ach.
    Var. galactophylla, Tuck.   Ocean Co.
P. stellaris, (L.,) Nyl.
    Var. tribacia, Fries.
    Var. hispida, Fries.
P. obscura, (Ehrh.,) Nyl.
    Var. endochrysea, Nyl.
    Var. adglutinata, Schrær.
P. pulverulenta, (Schreb.,) Nyl.

Pyxine, Fries.

P. cocoes, (Sw.)
    Var. sorediata, Tuck.

Umbilicaria, Hoffm.

U. Pennsylvanica, Hoffm.   Sussex Co.
U. pustulata, Hoffm.
U. Dillenii, Tuck.   Sussex Co.
U. Muhlenbergii, Ach.

Sticta, (Schreb.)

S. crocata, Ach.
S. quercizans, Ach.
S. pulmonaria, Ach.
S. amplissima, Mass.

Nephroma, Ach.

N. lævigatum, Ach.
N. tomentosum, (Hoffm.,) Kbr.
N. Helveticum, Schrær.

Peltigera, (Willd., Hoffm.,) Feé.

P. aphthosa, Hoffm.
P. canina, Hoffm.
Var. spuria, Ach.
P. polydactyla, Hoffm.
P. rufescens, Hoffm.
P. horizontalis, Hoffm.
P. venosa, Hoffm.

Pannaria, (Del.,) Tuckm.

P. lanuginosa, (Ach.,) Kbr.
P. lurida, Nyl.
P. microphylla, (Sw.,) Del.
P. leucosticta, Tuck.
P. molybdæa, Pers.
Var. cronia, Nyl.   Sussex Co.
P. nigra, (Huds.,) Nyl.
P. byssina, (Hoffm.,) Tuck.

Ephebe, Fries., Born.

E. pubescens, Fr.
E. minor, Willey in litt.

Collema, (Hoffm.,) Fries., Flot.

C. myriococcum, (Ach.,) Nyl.   Sussex Co.
C. pycnocarpum, Nyl.
C. cyrtaspris, Tuck.
C. microphyllum, Ach.
C. verruciforme, Nyl.
C. leptaleum, Tuck.
C. flaccidum, Ach.
C. nigrescens, (Huds.,) Ach.
C. ryssoleum, Tuck.
C. pulposum, Ach.
C. furvum, Ach., Nyl.

Leptogium, Fries., Nyl.

L. subtile, Nyl.
L. lacerum, (Sw.,) Fr.
L. pulchellum, Ach., Nyl.
L. tremelloides, Fr.
L. chloromelum, (Sw.,) Nyl.
L. myochroum, (Ehrh., Schrær.)
Var. saturninum, (Dicks.,) Tuck.

Hydrothysia, Russell.

    H. venosa, Russell.

Placodium, (DC.,) Naeg. & Hepp.

    P. cerinum, (Hedw.,) Naeg. & Hepp.

    P. aurantiacum, (Lightf.,) Naeg. & Hepp.

      Var. erythrellum, Ach.

    P. ferrugineum. (Huds..) Hepp.

      Var. nigricans, (Tuck.,) Fr.

    P. vittellinum, (Ehrb.,) Ach.

    P. cinnabarinum, (Ach.,) Anzi.

Lecanora, Ach.

    L. pallescens, Fr.

    L. athroocarpa, Duby, Nyl.

    L. rubina, Ach.   Sussex Co.

    L. tartarea, Ach.

    L. cinerea, (L.,) Sommerf.

    L. Bockii, (Fr.,) Th. & Fr.

    L. cervina, (Pers.,) Smf.

      Var. discreta, Fr.

      Var. privigna, Auctt.

    L. subfusca, Ach.

      Var. Hageni, Ach.

    L. pallida, Schrær.

    L. varia, Fr.

    L. orosthea, (Sm.,) Mudd.

    L. xanthophana, Nyl.

    L. muralis, (Schrær.)   Sussex Co.

Rinodina, Mass., Stitz.

    R. sophodes, (Ach.,) Mass.

      Var. confragosa, Nyl.

    R. Ascociscana, Tuck.   Sussex Co.

    R. constans, Nyl.

Pertussaria, DC.

    P. pertusa, Ach.

      Var. areolata, Fr.

    P. leioplaca, Ach.

    P. velata, Turn.

    P. multipuncta, (Sm.,) Nyl.

    P. pustulata, (Ach.,) Nyl.

    P. globularis, Ach.

Conotrema, Tuckm.
    C. urceolatum, Tuck.

Gyalecta, (Ach.,) Anzi.
    G. pineti, (Schrad.,) Fr.

Urceolaria, (Ach.,) Floerk.
    U. scruposa, Smf.

Stereocaulon, Schreb.
    S. tomentosum, Fr.
    S. paschale, Laur.　Sussex Co.
    S. denudatum, Fl.

Cladonia, Hoffm.
    C. papillaria, (Ehrh.,) Hoffm.
    C. pyxidata, Fr.
       Var. symphicarpa, Nyl.
    C. cariosa, Fl.
    C. turgida, Hoffm.
    C. fimbriata, Fr.
       Var. adspersa, Tuck.
    C. gracilis, Fr.
       Var. verticillata, Fr.
       Var. hybrida, Fr.
       Var. elongata, Fr.
       Var. symphicarpa, Tuck.
    C. cornuta, Fr.
    C. mitrula, Tuck.
    C. lepidota, Fr.
    C. furcata, Fl.
       Var. crispata, Fl.
       Var. racemosa, Fl.
       Var. subulata, Fl.
    C. squamosa, Hoffm.
       Var. delicata, Fr.
       Var. cæspiticia, Auctt.
    C. rangiferina, Hoffm.
       Var. sylvatica, Fl.
    C. degenerans, Fl.
    C. uncialis, Fr.　Ocean Co.
       Var. aduuca, Ach.　Ocean Co.
    C. cornucopioides, Fr.

C. macilenta, Hoffm.
C. cristatella, Tuck.

Bæomyces, Pers., DC., Nyl.

B. roseus, Pers.
B. æruginosus, Scop.

Biatora, Fries.

B. icteria, Mont.
B. Russellii, Tuck.   Sussex Co.
B. nigra, Tuck.
B. decolorans, (Hoffm.,) Fr.
B. viridescens, (Schrad.,) Fr.
B. vernalis, Fr.
B. parvifolia, Pers.
B. russula, (Ach.,) Mont.
B. sanguinea, Fr.
B. exigua, (Schrad.,) Ach.
B. uliginosa, (Schrad.,) Ach.
B. denigrata, Fr.
B. tricolor, With.
B. hypnophila, Turn.
B. cupreo-rosella, Nyl.   Sussex Co.
B. rubella, Fr.
    Var. suffusa, Tuck.
    Var. Schweinitzii, Tuck.
    Var. inundata, Fr.
B. umbrina, Ach.
B. chlorosticta, Tuck.   Ocean Co.
B. chlorantha, Tuck.
B. campestris, Fr.
B. fossarum, (Duf.,) Mont.
B. geophana, Nyl.
B. Resinæ, Fr.

Heterothecium, (Flot.,) Tuck.

H. sanguinarium, (L., Fl.,) Tuck.
H. pezizoideum, (Ach.,) Fl.

Lecidea, Ach., Fr.

L. contigua, Fr.
L. elæochroma, Tuck.
L. enteroleuca, Ach.

L. tessellina, Tuck.
L. spilota, Fr.

Buellia, (D. N.,) Tuck.

B. stellulata, Tayl.
B. parasema, (Ach.,) Kbr.
B. dialyta, Nyl.
B. myriocarpa, DC.
B. Schræreri, D. N.
B. vernicoma, Tuck.
B. petræa, (Fl.,) Tuck.
Var. Montagnei, Fl.
B. lactea, Mass.

Opegrapha, (Humb.,) Ach., Nyl.

O. varia, (Pers.,) Fr.
Var. rimalis, Fr.
O. vulgata, Ach., Nyl.
O. viridis, (Pers.,) Nyl.

Graphis, Ach., Nyl.

G. scripta, Ach.
Var. assimilis, Nyl.
Var. recta, Schræer.
G. elegans, (Sm.,) Ach.
G. dendritica, Ach.
G. scalpturata, Ach.

Arthonia, Ach., Nyl.

A. glaucescens, Nyl.
A. lecideella, Nyl.
A. astroidea, Ach.
A. spectabilis, Fl.
A. globosa, Tuckm., fide Willey in litt., Mar., 1875.

Mycoporum, Flot., Nyl.

M. pycnocarpum, Nyl.

Acolium, (Feé,) DN.

A. tigillare, (Ach.,) DN.

Calicium, Pers.

C. phæocephalum, (Turn.,) Turn. and Borr.
C. curtum, T. and B.
C. subtile, Fr.

C. fuscipes, Tuck.
C. roscidum, (Fl.,) Nyl.
   Var. roscidulum, Nyl.   Ocean Co.
C. byssaceum, Fr.
C. tubæforme, Tuck.

Endocarpon, Hedw., Fr.
   E. miniatum, (L.,) Schrær.
      Var. aquaticum, Schrær.
      Var. complicatum, Schrær.
   E. arboreum, Schweinitz.
   E. rufescens, Ach.
   E. hepaticum, Ach.

Trypethelium, Spreng., Ach.
   T. virens, Tuck.

Sagedia, (Mass., Kbr.,) Tuck.
   S. lactea, Kbr.
   S. oxyepora, (Nyl.,) Tuck.
   S. Cestrensis, Tuck.

Verrucaria, (Pers.,) Tuck.
   V. epigea, (Pers.,) Ach.

Pyrenula, (Ach., Naeg. & Hepp.,) Tuck.
   P. hyalospora, (Nyl.,) Tuck.
   P. glabrata, (Ach.,) Mass.
   P. nitida, Ach.
   P. lactea, (Mass.,) Tuck.
   P. punctiformis, (Ach.,) Naeg.
   P. thelæna, (Ach.,) Tuck.

FUNGI.

COMPILED BY MR. J. B. ELLIS.

AGARICINI.

Agaricus, L.
   Amanita.
      A. vaginatus, Bull.
      A. muscarius, L.

Lepiota.

    A. procerus, Scop.
    A. rachodes, Vitt.

Armillaria.

    A. melleus, Vahl.

Clitocybe.

    A. laccatus, Scop.
    A. trullissatus, Ell.

Pleurotus.

    A. ostreatus, Jacq.
    A. septicus, Fr.
    A. algidus, Fr.
    A. applicatus, Batsch.

Collybia.

    A. platyphyllus.
        Var. repens, Fr.
    A. confluens, Pers.
    A. conigenoides, Ell.
    A. acervatus, Fr.

Mycena.

    A. galericulatus, Scop.
    A. alcalinus, Fr.
    A. epipterygius, Scop.
    A. corticola, Schum.
    A. capillaris, Schum.

Omphalia.

    A. campanella, Batsch.
    A. fibula, Bull.

Entoloma.

    A. indigoferus, Ell.   (**A**. prunuloides, Fr.) (?)

Hebeloma.

    A. geophyllus, Bull.

Flammula.

    A. sapineus, Fr.

Nancoria.

    A. pediades, Fr.

Galera.

    A. tener, Schæff.

    A. hypnorum, Batsch.

Psalliota.

    A. campestris, L.

       Var. rufescens, Berk.

Stropharia.

    A. semiglobatus, Batsch.

Panæolus.

    A. campanulatus, L.

    A. atomatus, Fr.

    A. disseminatus, Fr.

Coprinus, Pers.

    C. comatus, Fr.

    C. atramentarius. Fr.

    C. fimetarius, Fr.

    C. niveus, Fr.

    C. micaceus, Fr.

    C. ephermerus, Fr.

Cortinarius, Fr.

    C. violaceus, Fr.

    C. cinnamomeus.

       Var. semisanguineus, Fr.

Paxillus, Fr.

    P. flavidus, Berk.

    P. atrotomentosus, Fr.

    P. pubescens, Ell.

Hygrophorus, Fr.

    H. virgineus, Fr.

    H. coccineus, Fr.

    H. miniatus, Fr.

    H. conicus, Fr.

Lactarius, Fr.

    L. torminosus, Fr.

    L. piperatus, Fr.

    L. vellereus, Fr.

L. deliciosus, Fr.
L. theiogalus, Fr.
L. volemus, Fr.
L. subdulcis, Fr.
L. indigo, Schw.

Russula, Fr.

R. furcata, Fr.
R. integra, Fr.
R. alutacea, Fr.

Cantharellus, Adans.

C. cibarius, Fr.
C. aurantiacus, Fr.
C. cinnabarinus, Schw.

Marasmius, Fr.

M. oreades, Fr.
M. scorodonius, Fr.
M. ramealis, Fr.
M. rotula, Fr.
M. androsaceus, Fr.
M. perforans, Fr.
M. epiphyllus, Fr.
M. praeacutus, Ell.
M. straminipes, Pk.
M. glabellus, Pk.
M. siccus, Schw.
M. cucullatus, Ell.

Lentinus, Fr.

L. Lecontei, Fr.
L. lepideus, Fr.

Panus, Fr.

P. strigosus, B. & C.
P. stypticus, Bull.

Trogia, Fr.

T. crispa, Fr.

Schizophyllum, Fr.

S. commune, Fr.

Lenzites, Fr.

    L. betulina, Fr.
    L. sepiaria, Fr.
    L. vialis, Pk.

POLYPOREI.

Boletus, Dill.

    B. luteus, L.
    B. granulatus, L.
    B. subtomentosus, L.
    B. edulis, Bull.
    B. scaber, Fr.
    B. felleus, Bull.
    B. strobilaceus, Berk.
    B. Frostii, Russ.
    B. Russelii, Frost.
    B. dichrous, Ell.

Polyporus, Fr.

    P. brumalis, Fr.
    P. Ellisii, Berk.
    P. Schweinitzii, Fr.
    P. perennis, Fr.
    P. parvulus, Klotszch.
    P. picipes, Fr.
    P. varius, Fr.
    P. lucidus, Fr.
    P. giganteus, Fr.
    P. sulfureus, Fr.
    P. epileucus, Fr.
    P. obtusus, Berk.
    P. adustus, Fr.
    P. applanatus, Fr.
    P. carneus, Nees.
    P. cupulaeformis, B. & Rav.
    P. pergamenus, Fr.
    P. igniarius, Fr.
    P. radiatus, Fr.
    P. hirsutus, Fr.
    P. velutinus, Fr.
    P. versicolor, Fr.
    P. abietinus, Fr.
    P. gilvus, Fr.

P. contiguus, Fr.
P. ferruginosus, Schrad.
P. xanthus, Fr.
P. nitidus, Fr.
P. incarnatus, Fr.
P. medulla-panis, Fr.
P. obducens, Fr.
P. vulgaris, Fr.
P. molluscus, Fr.
P. vaporarius, Fr.
P. aneirinus, Fr.
P. farinellus, Fr.
P. tenellus, Berk. & Cke.
P. xantholoma, Schw.

Trametes, Fr.

T. Pini, Fr.
T. sepium, Berk.
T. suaveolens, Fr.

Dædalea, Pers.

D. quercina, Fr.
D. confragosa, Pers.

Merulius, Hall.

M. tremellosus, Schrad.
M. corium, Fr.
M. aureus, Fr.
M. lacrymans, Fr.

Porothelium, Fr.

P. confusum, B. & Br.

Fistulina, Bull.

F. pallida, Berk. & Rav.

Favolus, Palis.

F. Europæus, Fr.

HYDNEI.

Hydnum, L.

H. repandum, L.
H. zonatum, Batsch.
H. ferrugineum, Fr.

H. adustum, Schw.
H. caput-medusae, Bull.
H. ochraceum, Pers.
H. Ellisii, Thüm.
H. farinaceum, Pers.
H. pallidum, C. & E.

Irpex, Fr.

I. cinnamomeus, Fr.
I. tulipifera, Schw.

Radulum, Fr.

R. orbiculare, Fr.

Phlebia, Fr.

P. merismoides, Fr.

Grandinia, Fr.

G. tabacina, C. & E.

Odontia, Fr.

O. fimbriata, Fr.
O. fusca, C. & E.

Kneiffia, Fr.

K. candidissima, B. & C.
K. setigera, Fr.

AURICULARINI.

Craterellus, Fr.

C. cornucopioides, Fr.

Thelephora, Ehr.

T. anthocephala, Fr.
T. terrestris, Fr.
T. palmata, Fr.
T. cristata, Fr.
T. (laciniata, P.?)
T pallida, Schw.
T. sebacea, Fr.
T. puteana, Schum.

Stereum, Fr.

S. purpureum, Fr.
S. hirsutum, Fr.

S. spadiceum, Fr.
S. acerinum, Fr.
   Var. nivosum, Fr.
S. papyrinum, Mont.
S. radiatum, Pk.

Hymenochæte, Lev.

H. rubiginosa, Lev.
H. tabacina, Lev.
H. Ellisii, B. & Cke.
H. corrugata, Berk.
H. spreta, Pk.
H. agglutinans, Ell.

Artocreas, Berk.

A. Micheneri, Berk.

Corticium, Fr.

C. arachnoideum, B. & C.
C. cinereum, Fr.
C. fumigatum, Thum.
C. rubrocanum, Thum.
C. læve, Fr.
C. calceum, Fr.
C. giganteum, Fr.
C. incarnatum, Fr.
C. molle, B. & C.
C. subrepandum, Berk. & Cooke.
C. lilacino-fuscum, B. & C.
C. glabrum, B. & C.
C. brunneolum, B. & C.
C. subgiganteum, B. & C.
C. echinospoum, Ell.
C. ochroleucum, Fr.
   Var. spumeum, B. & Rav.
C. fusisporum, C. & E.
C. effuscatum, C. & E.
C. polygonium, Fr.
C. colliculosum, B. & C.

Exobasidium, Wor.

E. Vaccinii, Woron.
   Var. discoidea, Ell.
   Var. Andromedæ, Pk.

Cyphella, Fr.

C. fulva, B. & Rav.

Solenia, Pers.

S. ochracea, Hoff.
S. candida, Hoff.
S. fasciculata, Hoff.

CLAVARIEI.

Clavaria, Vaill.

C. cristata, Holmsk.
C. inequalis, Müll.
C. clavata, Pk.
C. mucida, Pers.

Calocera, Fr.

C. cornea, Fr.

Typhula, Fr.

T. muscicola, Fr.

Pistillaria, Fr.

P. micans, Fr.
P. clavulata, Ell.

TREMELLINI.

Tremella, Dill.

T. foliacea, P.
T. albida, Huds.
T. stipitata, Pk.

Exidia, Pers.

E. glandulosa, Fr.

Hirneola, Fr.

H. Auricula-Judæ, Fr.

Næmatelia, Fr.

N. nucleata, Schw.
N. encephala, Fr.

Dacryomyces, Nees.

D. deliquescens, Duby.
D. stillatus, Nees.

Hymenula, Fr.
>   **H.** fumosa, C. & E.

Hormomyces, Bon.
>   **H.** aurautiacus, Bon.

Ditiola, Fr.
>   (**D.** radicata, Fr.)?

HYPOGÆI.

Rhizopogon, Fr.
>   **R.** rubescens, Tul.

PHALLOIDEI.

Phallus, Mich.
>   **P.** impudicus, L.

Corynites, B. & C.
>   **C.** Ravenelii, B. & C., (Elizabeth, Tom's River and Fort Lee, W.
>   R. Gerard.)

TRICHOGASTRES.

Geaster, Mich.
>   **G.** hygrometricus, Pers.
>   **G.** mammosus, Chev.
>   **G.** minimus, Schw.

Bovista, Pers.
>   **B.** subterranea, Pk.

Lycoperdon, Tourn.
>   **L.** cyathiforme, Bosc.
>   **L.** pusillum, Fr.
>   **L.** gemmatum, Fr.
>   **L.** Wrightii, B. & C.
>   **L.** pyriforme, Schæff.

Scleroderma, Pers.
>   **S.** bovista, Fr.
>   **S.** vulgare, Fr.

Arachnion, Schw.
>   **A.** album, Schw.

22

Polysaccum, D. C.

    P. Pisocarpium, Fr.

MYXOGASTRES.

Lycogala, Mich.

    L. epidendrum, Fr.

Reticularia, Bull.

    R. atra, Fr.

    R. umbrina, Fr.

Licea, Schrad.

    L. applanata, Berk.

Fuligo, Hall.

    F. varians, Sommf.

Leocarpus, Link.

    L. fragilis, (Dicks.,) Rost.

Chondrioderma, Rost.

    C. floriforme, (Bull.,) Rost.

Didymium, Schrad.

    D. tigrinum, Fr.

    D. squamulosum, A. & S.

    D. xanthopus, Fr.

    D. Michelii, Lib.

    D. cinereum, Fr.

Angioridium, Grev.

    A. sinuosum, Grev.

Badhamia, Berk.

    B. hyalina, Berk.

    B. penetrans, C. & E.

Enteridium, Ehr.

    E. olivaceum, Ehr.

Diachea, Fr.

    D. elegans, Fr.

Stemonitis, Gled.

    S. ferruginea, Ehrb.

    S. fusca, Roth.

    S. confluens, C. & E.

Dictydium, Schrad.

    D. umbilicatum, Schrad.

Cribraria, Pers.

    C. intricata, Schrad.

Arcyria, Hill.

    A. punicea, Pers.

    A. incarnata, Pers.

    A. cinerea, Schum.

Ophiotheca,

    O. umbrina, Berk.

Trichia, Hall.

    T. rubiformis, Pers.

    T. fallax, Pers.

    T. scabra, Rost.

    T. chrysosperma, DC.

    T. varia, Pers.

    T. serpula, Pers.

Hemiarcyria, Rost.

    H. clavata, Pers.

Perichæna, Fr.

    P. depressa, Lib.

NIDULARIACEI.

Cyathus, Hall.

    C. vernicosus, D. C.

Crucibulum, Tul.

    C. vulgare, Tul.

Sphærobolus, Tode.

    S. stellatus, Tode.

SPHÆRONEMEI.

Coniothyrium, Corda.

    C. subtile, Cda.

    C. herbarum, C. & E.

Leptostroma, Fr.

    L. litigiosum, Desm.

    L. caricinum, Fr.

    L. petiolorum, C. & E.

Phoma, Fr.

    P. concentricum.
    P. nebulosum, Berk.
    P. longissimum, Berk.
    P. fibricola, Berk.
    P. acuum, C. & E.
    P. uvicola, Berk. & Curt.
    P. ampelina, B. & C.
    P. consorta, C. & E.

Cryptosporium, Kze.

    C. epiphyllum, C. & E.
    C. Loniceræ, C. & E.
    C. Nyssæ, C. & E.
    C. Solidaginis, C. & E.

Sphæronema, Fr.

    S. acerinum, Pk.
    S. corneum, C. & E.
    S. macrosporum, B. & C.
    S. pruinosum, B. & C.
    S. rufum, Fr.
    S. subcorticale, C. & E.
    S. hystricinum, Ell.
    S. stellatum, Ell.
    S. hispidulum, Ell.
    S. clethrincola, Ell.
    S. sphæroideum, Ell.
    S. capillare, E. & H.

Sphæropsis, Mont.

    S. malorum, Berk.
    S. Alni, C. & E.
    S. Ampelopsidis, C. & E.
    S. clethræcola, C. & E.
    S. Cydoniae, C. & E.
    S. diatrypeum, C. & E.
    S. fibriseda, C. & E.
    S. Ilicicola, C. & E.
    S. lanceolata, C. & E.
    S. opaca, C. & E.
    S. phacidioides, C. & E.
    S. pinastri, C. & E.
    S. punctum, C. & E.

S. ribicola, C. & E.
S. rubicola, C. & E.
S. Rosarum, C. & E.
S. Sassafras, C. & E.
S. Sumachi, C. & E.
S. valsoidea, C. & E.

Diplodia, Fr.

D. herbarum, Lev.
D. viticola, Desm.
D. ilicicola, Desm.
D. asclepiadea, C. & E.
D. atramentaria, C. & E.
D. decorticata, C. & E.
D. glandicola, C. & E.
D. hibiscina, C. &. E.
D. hyalospora, C. & E.
D. longispora, C. & E.
D. maura, C. & E.
D. moricola, C. & E.
D. radicina, C. & E.
D. thyoidea, C. & E.

Hendersonia, Mont.

H. sarmentorum, West.
H. Cydoniæ, C. & E.
H. anomala, C. & E.
H. collapsa, C. & E.
H. delicatula, C. E.
H. lophiostoma, C. & E.
H. thyoides, C. & E.
H. trimera, Cooke.

Vermicularia, Tode.

V. atramentaria, B. & Br.
V. dematium, Fr.
V. compacta, C. & E.
V. venturioidea, C. & E.

Discosia, Fr.

D. artocreas, Fr.
D. Podisomæ, C. & E.

Pilidium, Kze.

P. quercinum, Cke.

Septoria, Fr.

> S. graminum, Desm.
> S. Polygonorum, Desm.
> S. ilicifolia, C. & E.
> S. Liquidambaris, C. & E.
> S. kalmicola, B. & C.
> S. tenella, C. & E.
> S. Œnotheræ, West.
> S. phlyctænoides, B. & C.
> S. Lactucæ, Pass.

Phyllostica, Pers.

> P. acericola, C. & E.
> P. Myricæ, Cke.

Excipula, Fr.

> E. strigosa, Fr.
> E. microspora, C. & E.

Dinemasporium, Lev.

> D. graminum, Lev.
> D. patellum, C. & E.
> D. Robiniæ, Ger.

Myxormia, B. & Br.

> M. convexula, C. & E.

Asteroma, D. C.

> A. Rosæ, D. C.

Micropera, Lev.

> M. drupacearum, Lev.

MELANCONIEI.

Melanconium, Link.

> M. bicolor, Nees.
> M. magnum, Berk.
> M. oblongum, Berk.

Stilbospora, Pers.

> S. ovata, Pers.

Coryneum, Nees.

> C. Kunzei, Cda.

Pestalozzia, De Not.

    P. monochaetoidea, Sacc. & Ell.

    P. conigena, Lev.

    P. clavata, C. & E.

    P. hysteriiformis, B. & C.

    P. pezizoides, De Not.

    P. stellata, B. & C.

    P. unicornis, C. & E.

Nemaspora, Fr.

    N. crocea, Pers.

Myxosporium, De Not.

    M. nitidum, Berk.

TORULACEI.

Torula, Pers.

    T. herbarum, Lk.

    T. opaca, Cke.

    T. binale, C. & E.

    T. bigemina, C. & E.

    T. sphæriæformis, C. & E.

Speira, Corda.

    S. punctulata, C. & E.

Septonema, Corda.

    S. spilomeum, Berk.

    S. bicolor, Pk.

    S. tabacinum, E. & H.

    S. rude, Sacc.

    S. toruloides, C. & E.

Sporidesmium, Link.

    S. polymorphum, Cda.

    S. antiquum, Cda.

    S. lepraria, B. & Br.

    S. rude, Ell.

Tetraploa, B. & Br.

    T. Ellisii, Cke.

Gymnosporium, Corda.

    G. Arundinis, Cda.

PUCCINIAEI.

Phragmidium, Link.

P. mucronatum, Lk.
P. obtusum, Lk.
P. speciosum, Fr.

Puccinia, Pers.

P. graminis, Pers.
P. arundinacea, Hedw.
P. striola, Lk.
P. coronata, Cda.
P. Polygonorum, Lk.
P. Menthæ, Pers.
P. Compositarum, Sch.
P. Galiorum, Lk.
P. noli-tangeris, Cda.
P. Violarum, Lk.
P. Epilobii, D. C.
P. Circaeæ, Pers.
P. Prunorum, Lk.
P. Caricis, D. C.
P. Helianthi, Schw.
P. Smilacis, Schw.
P. Sorghi, Schw.
P. Xanthii, Schw.
P. Ellisii, Thum.

Gymnosporangium, D. C.

G. macropus, Schw.
G. Ellisii, Berk.
G. biseptatum, Ell.
G. clavipes, C. & P.
G. (foliicolum, Cda.) (?)   On foliage of white cedar.

CÆOMACEI.

Ustilago, Link.

U. carbo, Tul.
U. Maydis, Cda.
U. Junci, Schw.
U. utriculosa, Tul.
U. Syntherismæ, Schw.

Tubercinia, Fr.

T. Scabies, Berk.

Urocystis, Rabh.

U. cepulæ, Frost.

Uromyces, Lev.

U. appendiculatus, Lev.

U. Polygoni, Fckl.

U. Spermacocis, Schw.

U. Lespedezæ, Schw.

U. Euphorbiæ, C. & P.

U. Asclepiadis, Cke.

U. Hyperici, Schw.

U. Caladii, Schw.

Coleosporium, Lev.

C. Solidaginis, (Schw.,) Thum.

C. miniatum, Pers.

Melampsora, Cast.

M. Salicina, Lev.

M. Populina, Lev.

Cystopus, D. Bary.

C. candidus, Lev.

C. cubicus, Strauss.

C. Portulaccæ, De By.

Uredo, Lev.

U. luminata, Schw.

U. (Vacciniorum, Pers.) (?)

U. pyrolata, Kornicke.

Æ CIDIA CEI.

Rœstelia, Reb.

R. cancellata, Reb.

R. cornuta, Tul.

R. lacerata, Tul.

R. transformans, Ell.

R. aurantiaca, Pk.

R. Botryapites, Schw.

Peridermium, Chev.

P. Pini, Chev.

P. cerebrum, Pk.

Æcidium, Pers.

    Æ. Euporbiæ, Pers.

    Æ. Berberidis, Pers.

    Æ. crassum, Pers.

    Æ. Ranunculacearum, DC.

    Æ. Compositarum, Mart.

    Æ. Violæ, Schum.

    Æ. Myricatum, Schw.

    Æ. Caladii, Schw.

ISARIACEI.

Ceratium, A. & S.

    C. hydnoides, A. & S.

STILBACEI.

Stilbum, Tode.

    S. erythrocephalum, Ditt.

    S. atrocephalum, Ell.

    S. Rhoidis, B. & C.

    S. parvulum, C. & E.

    S. æruginosum, Desm,

    S. Smaragdinum, A. & S.

Volutella, Fr.

    V. flexuosa, C. & E.

    V. (hyacinthorum, Berk.) (?)

Tubercularia, Tode.

    T. nigricans, Lk.

    T. vulgaris, Tode.

Fusarium, Link.

    F. roseum, Lk.

    F. heterosporium, Nees.

    F. lateritium, Nees.

    F. diplosporum, C. & E.

    F. thujinum, Ell.

    F. miniatum, Sacc.

Myrothecium, Tode.

    M. verrucaria, A. & S.

Epicoccum, Link.

    E. scabrum, Cda.

    E. sphærospermum, Berk.

    E. Duriæanum, Mont.

Illosporium, Mont.

    I. pallidum, Cke.

    I. coccinellum, Cke.

Ægerita, Pers.

    Æ. candida, Pers.

DEMATIEI.

Arthrobotryum, Ces.

    A. robustum, C. & E.

Arthrosporium, Sacc.

    A. compositum, Ell.

Dendryphium, Corda.

    D. Ellisii, Cke.

    D. Harknessii, Ell.

Periconia, Corda.

    P. Azaleæ, Pk.

Sporocybe, Fr.

    S. byssoides, Fr.

Helminthosporium, Link.

    H. arctæsporum, C. & E.

    H. brachytrichum, C. & E.

    H. inconspicuum, C. & E.

    H. inflatum, B. & Bav.

    H. interseminatum, B. & Rav.

    H. leptotrichum, C. & E.

    H. macrocarpon, Grev.

    H. persistens, Cke.

    H. subopacum, C. & E.

Macrosporium, Fr.

    M. chartarum, Pk.

    M. Porri, C. & E.

    M. abruptum, C. & E.

    M. atrichum, C. & E.

    M. fasciculatum, C. & E.

    M. nigrellum, C. & E.

    M. inquinams, C. & E.

    M. leptotrichum, C. & E.

Mystrosporium, Corda.

    M. orbiculare, C. & E.

    M. aterrimum, B. & C.

Acrothecium, Corda.

    A. obovatum, C. & E.

Septosporium, Corda.

    S. maculatum, C. & E.

    S. prælongum, Sacc.

    S. velutinum, C. & E.

Helicoma, Corda.

    H. Mulleri, Cda.

Helicosporium, Nees.

    H. olivaceum, Pk.

    H. ellipticum, Pk.

    H. auratum, Ell.

    H. thysanophorum, E. & H.

Polythrincium, Kze.

    P. Trifolii, Kze.

Cladosporium, Link.

    C. herbarum, Lk.

    C. epiphyllum, Nees.

    C. delectum, C. & E.

    C. graminum, Lk.

Clasterisporium, Schw.

    C. subulatum, Cke.

    C. herculeum, Ell.

MUCEDINES.

Aspergillus, Mich.

    A. glaucus, Lk.

Rhinotrichum, Corda.

    R. ramosissimum, B. & C.

    R. Curtisii, Berk.

    R. macrosporum, Farlow.

    R. repens, Preuss.

Botrytis, Mich.

    B. geniculata, Cda.

    B. atrofumosa, C. & E.

B. atroviridis, C. & E.
B. friliginosa, C. & E.
B. acinorum, Pers.

Polyactis, Link.

P. vulgaris, Lk.
P. streptothrix, C. & E.

Penicillium, Link.

P. crustaceum, Fr.
P. repens, C. & E.

Oidium, Link.

O. monilioides, Lk.
O. megalosporum, Berk.
O. simile, Berk.

Monilia, Hill.

M. Martinii, Ell. & Sacc.
   Var. fructigena,

Dactylium, Nees.

D. roseum, Berk.

Sporotrichum, Link.

S. sulfureum, Grev.
S. æruginosum, Schw.

Zygodesmus, Corda.

Z. fuscus, Cda.
Z. bicolor, C. & E.
Z. rubiginosus, Pk.
Z. chlorochactes, Ell.
Z. olivasceus, B. & C.

Myxotrichum, Kze.

M. ochraceum, B. & Br.

Gonytrichum, Nees.

G. fulvum, Ell.

Menispora, Pers.

M. ciliata, Cda.
M. glauconigra, C. & E.

Campsotrichum, Ehrb.

C. flagellum, C. & E.

SEPEDONIEI.

Sepedonium, Link.

    S. chrysospermum, Lk.

Fusisporium, Link.

    F. aurantiacum, Lk.

    F. episphericum, C. & E.

    F. pallido-roseum, Cke.

    F. Andropogonis, Cke.

Epochnium, Link.

    E. macrosporoideum, Berk.

Ellisiella, Sacc.

    E. caudata, Sacc.

Colletotrichum, Corda.

    C. lineola, Cda.

AUTENNARIEI.

Zasmidium, Fr.

    Z. cellare, Fr.

MUCORINI.

Mucor, Mich.

    M. phycomyces, Berk.

    M. caninus, Pers.

Hydrophora, Tode.

    H. stercorea, Tode.

Sporodinia, Link.

    S. dichotoma, Cda.

Syzigites, Ehrb.

    S. megalocarpus, Ehr.

Endogone, Link.

    E. pisiformis, Lk.

PERISPORIACEI.

Sphærotheca, Lev.

    S. Castagnei, Lev.

Phyllactinia, Lev.

    P. guttata, Lev.

Uncinula, Lev.

   U. adunca, Lev.

Podosphæra, Kze.

   P. Kunzei, Cda.

Microsphæra, Lev.

   M. Hedwigii, Lev.
   M. extensa, C. & P.
   M. penicillata, Lev.

Erysiphe, Hedw.

   E. lamprocarpa, Lev.
   E. Martii. Lk.

Chætomium, Kze.

   C. elatum, Kze.
   C. olivaceum, C. & E.
   C. sphærospermum, C. & E.
   C. chartarum, Ehrb.

Eurotium, Link.

   E. herbariorum, Lk.

ELVELLACEI.

Morchella, Dill.

   M. esculenta, Pers.

Helvella, Linn.

   H. lacunosa, Afz.

Mitrula, Fr.

   M. paludosa, Fr.

Leotia, Hill.

   L. lubrica, Pers.

Vibrissea, Fr.

   V. truncorum, Fr.

Geoglossum, Pers.

   G. glabrum, Pers.
   G. hirsutum, Pers.
   G. rufum, Schu.
   G. luteum, Pk.

Peziza, Linn.

P. acetabulum, L.
P. macropus, Pers.
P. repanda, Wahl.
P. vesciculosa, Bull.
P. cerea, Sow.
P. nebulosa, Cke.
P. constellatio, B. & Br.
P. orthotricha, C. & E.
P. melastoma, Sow.
P. stygia, B. & C.
P. hemispherica, Wigg.
P. scutellata, Linn.
P. trechispora, B. & Br.
P. fuscidula, C. & E.
P. virginea, Batsch.
P. nivea, Fr.
P. albopileata, Cke.
P. virginella, Cke.
P. calycina, Schum.
P. bicolor, Bull.
P. lachnoderma, Berk.
P. pollinaria, Cke.
P. chameleontina, Pk.
P. variecolor, Fr.
P. hyalina, Pers.
P. miniopsis, Ell.
P. acerina, C. & E.
P. cupressina, Batsch.
P. echinulata, Awd.
P. myricacea, Pk.
P. pulverulenta, Lib.
P. luteodisca, Pk.
P. Osmundæ, C. & E.
P. theioidea, C. & E.
P. culcitella, C. & E.
P. aurelia, Pers.
P. fusoa, Pers.
P. subiculata, Schw.
P. sanguinea, Pers.
P. radiocincta, Cke.
P. pruinata, Schw.
P. scutula, Pers.

P. cyathoidea, Bull.
P. culmicola, Desm.
P. simulans, Ell.
P. incondita, Ell.
P. nyssægena, Ell.
P. gracilipes, Cke.
P. nigrescens, Cke.
P. fumosella, C. & E.
P. Œnotheræ, C. & E.
P. aquifolia, C. & E.
P. introviridis, C. & E.
P. stictoidea, C. & E.
P. astericola, C. & E.
P. atrata, Pers.
P. atrovirens, Pers.
P. atrocinerea, Cke.
P. subatra, C. & P.
P. cinerea, Batsch.
P. rhaphidospora, Ell.
P. introspecta, Cke.
P. diaphanula, Cke.
P. erigeronata, Cke.
P. atriella, Cke.
P. cervinula, Cke.
P. mauriatra, C. & E.
P. denigrata, Kze.
P. melatephra, Lasch.
P. protrusa, B. & C.
P. Pteridis, A. & S.
P. vinosa, A. & S.
P. rubella, Pers.
P. paulopuncta, C. & E.
P. regalis, C. & E.
P. tenella, C. & E.
P. exigua, Cke.
P. coccinella, Somm.
P. vulgaris, Fr.
P. aureofulva, Cke.
P. corneola, C. & P.
P. macrospora, Fckl.
P. resinæ, Fr.

Chlorosplenium, Fr.

C. æruginosum, Fr.

C. Schweinitzii, Fr.
C. epimyces, Cke.

Dermatea, Fr.
D. furfuracea, Fr.
D. tabacina, Cke.
D. olivacea, Ell.
D. tetraspora, Ell.
D. lobata, Ell.
D. fascicularis, A. & S.
D. carnea, C. & E.
Var. pallida, Ell.
D. cucurbitaria, Cke.
D. Kalmiae, Pk.

Helotium, Tode.
H. subtile, Fr.
H. virgultorum, Fr.
H. citrinum, Fr.
Var. confluens, Schw.
H. herbarum, Fr.
H. epiphyllum, Fr.
H. albovirens, Cke.
H. album, Schum.
H. aurantiacum, Cke.
H. gracile, C. & P.
H. naviculasporum, Ell.
H. renisporum, Ell.

Patellaria, Fr.
P. atrata, Fr.
P. rhabarbarina, Berk.
P. ligniota, Fr.
P. aureo-coccinea, B. & C.
P. clavata, Ell.
P. connivens, Fr.
P. cylindrospora, Ell.
P. ferruginea, C. & E.
P. fuscoatra, Rehm.
P. gnaphaliana, C. & E.
P. imperfecta, Ell.
P. subsidua, C. & E.
P. tuberculosa, Ell.

Tympanis, Tode.

    T. conspera, Fr.
    T. alnea, Pers.

Cenangium, Fr.

    C. pulveraceum, Fr.
    C. turgidum, Schw.
    C. Cephalanthi, Schw.
    C. acuum, C. & P.
    C. triangulare, Schw.
    C. urceolatum, Ell.

Ascobolus, Tode.

    A. furfuraceus. Pers.
    A. Leveillei, Boud.
      Var. Americana, C. & E.
    A. ciliatus, Schmidt & Kze.

Bulgaria, Fr.

    B. inquinans, Fr.
    B. purpurea, Fckl.
    B. sarcoides, Fr.

Agyrium, Fr.

    A. rufum, Fr.
    A. herbarum, Fr.
    A. sexdecemsporum, Fckl.

Stictis, Pers.

    S. radiata, Pers.
    S. pustulata, Ell.
    S. sphaeroboloidea, Schw.
    S. Sesleriae, Lib.
    S. dryophila, C. & E.
    S. fimbriata, Schw.
    S. quercifolia, C. & E.
    S. stigma, C. & E.
    S. stereicola, B. & C.
    S. linearis, C. & E.

Propolis, Corda.

    P. grisea, C. & E.
    P. lobata, C. & E.
    P. versicolor, Fr.
    P. conorum, Ell.
    P. Leonis, Tul.

Ascomyces, M. & D.

    A. deformans, Berk.

    A. bullata, Berk.

    A. anomalus, E. & H.

TUBERACEI.

Elaphomyces, Nees.

    E. granulatus, Fr.

PHACIDIACEI.

Phacidium, Fr.

    P. dentatum, Fr.

    P. Trifolii, Boud.

    P. sphaeroideum, C. & E.

Rhytisma, Fr.

    R. acerinum, Fr.

    R. decolorans, Schw.

Triblidium, Reb.

    T. insculptum, Cke.

Hysterium, Tode.

    H. subrugosum, C. & E.

    H. pulicare, Pers.

    H. Rousselii, De Not.

    H. complanatum, Duby.

    H. Smilacis, Schw.

    H. Cookeanum, Ger.

    H. flexuosum, Schw.

    H. Gerardi, C. & P.

    H. viticolum, C. & P.

    H. hyalinum, C. & P.

    H. ellipticum, Fr.

    H. Mori, Schw.

    H. Nova-Cæsariense, Ell.

    H. gloniopsis, Ger.

Glonium, Schw.

    G. lineare, Fr.

    G. parvulum, Ger.

    G; simulans, Ger.

    G. stellatum, Muhl.

Angelina, Duby.

    A. rufescens, (Schw.,) Duby.

Hypoderma, D. C.

    H. virgultorum, D. C.

    H. commune, Fr.

Lophodermium, Chev.

    L. maculare, Fr.

    L. pinastri, Schrad.

    L. exaridum, C. & P.

    L. arundinaceum, Schrad.

Ailographum, Lib.

    A. vagum, Desm.

    A. culmigenum, Ell.

Lophium, Fr.

    L. mytilinum, Fr.

SPHAERIACEI.

Torrubia, Lev.

    T. militaris, Fr.

    T. ophioglossoides, Tul.

Epichloe, Fr.

    E. typhina, Pers.

    E. Hypoxylon, Pk.

Hypocrea, Fr.

    H. rufa, Fr.

    H. contorta, Schw.

    H. consimilis, Ell.

    H. chlorospora, B. & C.

    H. armeniaca, B. & C.

    H. olivacea, C. & E.

    H. citrina, Fr.

    H. Geoglossi, Ell.

Nectria, Fr.

    N. cinnabarina, Fr.

    N. pulicaris, Fr.

    N. coccinea, Fr.

    N. cucurbitula, Tode.

N. punicea, Kze.
N. sanguinea, Fr.
N. episphaeria, Fr.
N. aurigera, B. & Rav.
N. aureofulva, C. & E.
N. microspora, C. & E.
N. depauperata, Cke.
N. Brassicae, Ell. and Sacc.
N. vulpina, Cke.

Xylaria, Fr.

X. corniformis, Mont.
X. Hypoxylon, Grev.
X. filiformis, A. & S.

Ustulina, Tul.

U. vulgaris, Tul.

Nummularia, Tul.

N. Bulliardi, Tul.

Hypoxylon, Bull.

H. concentricum, Grev.
H. coccineum, Bull.
H. rubiginosum, Fr.
H. fuscum, Pers.
H. punctulatum, B. & Rav.
H. epiphloeum, B. & C.
H. serpens, Fr.

Eutypa, Tul.

E. lata, Tul.
E. Acharii, Tul.
E. spinosa, Tul.
E. leioplaca, Fr.
E. velutina, Sacc.

Melogramma, Tul.

M. fuliginosum, Ell.

Dothidea, Fr.

D. Trifolii, Fr.
D. Junci, Fr.
D. Pteridis, Fr.
D. flabella, Schw.

D. ribesia, Pers.
D. filicina, Fr.
D. Solidaginis, Schw.
D. Heliopsidis, Schw.
D. Muhlenbergii, Ell.
D. excavata, C. & E.
D. moricola. C. & E.
D. tetraspora, B. & Br.

Diatrype, Fr.

D. quercina, Pers.
D. moriformis, C. & P.
D. microspora, Ell.
D. discoidea, C. & P.
D. stigma, Fr.
D. disciformis, Hoff.
D. hystrix, Tode.
D. dryophila, Curr.
    Var. minor, Curr.
D. cincta, B. & Br.
D. quadrata, Schw.
D. grandinea, B. & Rav.
D. platystoma, Schw.
D. hypophloea, B. & C.
D. microplaca, B. & C.
D. collariata, C. & E.
D. rhuina, C. and E.
D. olivacea, C. & E.
D. irregularis, C. & E.
D. fibritecta, C. & E.
D. Duriaei, Mont.
D. albo-pruinosa, Schw.
D. anomala, Pk.

Melanconis, Tul.

M. sigmoidea, C. & E.

Valsa, Fr.

V. stellulata, Fr.
V. leucostoma, Fr.
V. ceratophora, Tul.
V. Abietis, Fr.
V. ambiens, Fr.
V. salicina, Fr.

V. tetraploa, B. & C.
V. pulchella, Fr.
V. pulchelloidea, C. & E.
V. venusta, Ell.
V. quaternata, Pers.
V. leiphaemia, Fr.
V. thelebola, Fr.
V. profusa, Fr.
V. Vitis. Schw.
V. ventriosa, C. & E.
V. subcuticularis, C. & E.
V. sociata, C. & E.
V. rugiella, C. & E.
V. rhuiphila, C. & E.
V. praestans, B. & C.
V. Americana, B. & C.
V. phomaspora, C. & E.
V. personata, C. & E.
V. Pennsylvanica, B. & C.
V. pauperata, C. & E.
V. paulula, C. & E.
V. parasitica, C. & E.
V. ocularia, C. & E.
V. obtecta, C. & E.
V. nigrofacta, C. & E.
V. Myricae, C. & E.
V. myinda, C. & E.
V. multiplex, C. & E.
V. Maclurae, C. & E.
V. Liquidambaris, Schw.
V. ligustrina, Cke.
V. laurina, C. & E.
V. juglandina, C. & E.
V. inconspicua, C. & E.
V. obscura, Pk.
V. femoralis, Pk.
V. tuberculosa, Ell.
V aouloans, Schw.
V. rufescens, Schw.

Cucurbitaria, Gray.

C. elongata, Grev.
C. Comptoniae, C. & E.
C. morbosa, Schw.

Massaria, De Not.

 M. vomitoria, B. & C.
 M. bufonia, Tul.
 M. epileuca, B. & C.

Lophiostoma, De Not.

 L. scelestum, C. & E.
 L. tingens, Ell.

Sphæria, Hall.

 S. aquila, Fr.
 S. pezizula, B. & C.
 S. solaris, C. & E.
 S. subiculata, Schw.
 S. ligniaria, Grev.
 S. abietina, Fckl.
 S. scopula, C. & P.
 S. hirsuta, Fr.
 S. canescens, Pers.
 S. ovina, Pers.
 S. xylariaespora, C. & E.
 S. atrobarba, C. & E.
 S. manmiformis, Pers.
 S. obducens, Fr.
 S. millegrana, Schw.
 S. cirrhosa, Pers.
 S. avocetta, C. & E.
 S. melanotes, B. & Br.
 S. thuriodonta, C. & E.
 S. sepelibilis, B. & C.
 S. segna, C. & E.
 S. arctaespora, C. & E.
 S. atrograna, C. & E.
 S. biglobosa, C. & E.
 S. bispherica, C. & E.
 S. melanostigma, C. & E.
 S. rugulosa, Fckl.
 S. albocincta, C. & E.
 S. paecilostoma, B. & Br.
 S. vetusta, Ell.
 S. caminata, C. & E.
 S. bacillata, Cke.
 S. parallela, Fr.
 S. luteobasis, Ell.

S. picacea, C. & E.
S. diaphana, C. & E.
S. phomopsis, C. & E.
S. aliquanta, C. & E.
S. vexata, C. & E.
S. secreta, C. & E.
S. castanella, C. & E.
S. microtheca, C. & E.
S. Catariæ, C. & E.
S. fissicula, C. & E.
S. verbascicola, Schu.
S. sublanosa, Cke.
S. adelphica, Cke.
S. Ilicis, Schleich.
S. latebrosa, Ell.
S. cryptica, Niessl.
S. gallophila, Ell.
S. spiculosa, Fr.
S. orthoceras, Fr.
S. euspina, C. & E.
S. Phaseolorum, C. & E.
S. salviæcola, C. & E.
S. subexserta, C. & E.
S. Desmodii, Pk.
S. tumulata, C. & E.
S. calvescens, Fr.
S. doliolum, Pers.
S. culmifraga, Fr.
S. Ogilviensis, B. & Br.
S. orthogramma, B. & C.
S. Bokoniæ, C. & E.
S. aulica, C. & E.
S. Tephrosiæ, C. & E.
S. comatella, C. & E.
S. distributa, C. & E.
S. consessa, C. & E.
S. clavigera, C. & E.
S. dissiliens, C. & E.
S. Virginica, C. & E.
S. Bidwellii, Ell.
S. atriella, C. & E.
S. cariosa, C. & E.
S. Hendersonia, Ell.

S. pachyascus, C. & E.
S. squamata, C. & E.
S. dumetorum, Niessl.
S. leiostega, Ell.
S. Eckfeldtii, Ell.
S. inflata, Ell.
S. soluta, C. & E.
S. subcutanea, C. & E.
S. rubicunda, Niessl.
S. herbarum, Pers.
S. denotata, C. & E.
S. barbirostris, Dufour.
S. ambleia, C. & E.
S. Clavariæ, Awd.
S. flabelliformis, Schw.
S. Kalmiarum, Schw.
S. Andromedæ, Schw.
S. Sarraceniæ, Schw.

Ceratostoma, Fr.

C. fallax, Cke. & Sacc.

Meliola, Fr.

M. nidulans, Schw.
M. fenestrata, C. & E.
M. maculosa, Ell.

Venturia, De Not.

V. barbula, B. & Br.
 Var. foliicola, Ell.
V. cupressina, Rehm.
V. ditricha, Fr.
V. Kalmiæ, Pk.
V. pulchella, C. & P.

Sphærella, De Not.

S. myriadea, DC.
S. Gaultheriæ, C. & E.
S. maculæformis, Pers.
S. nyssæcola, Cke.
S. pardalota, C. & E.
S. punctiformis, Pers.
S. sentina, Fr.
S. hypericina, Ell.

Gnomonia, (Rabh.)
> **G.** Myricæ. C. & E.
> **G.** setacea, Pers.

Microthyrium, Desm.
> **M.** Smilacis, De Not.

Dichæna, Fr.
> **D.** quercina, Fr.
> **D.** strumosa, Fr.

Capnodium, Mont.
> **C.** elongatum, Berk. & Desm.
> **C.** australis, Mont.

The foregoing list of about 950 species of New Jersey Fungi, though not professing to be a complete enumeration of all the species to be found within the limits of the State, is the best that can at present be given.

The list of the Agaricini especially is very meagre, embracing mostly only the commonest species. It is probable that of this family alone there are, within the limits of the State, from the mountainous districts of the north to the low plains of the south, at a moderate estimate, at least 200 species.

All the species enumerated have been collected in the immediate vicinity of Newfield, and as far as the ascomycetous fungi are concerned, the list probably includes the bulk of the species which grow in that vicinity.

In his Synopsis of North American Fungi, Schweinitz mentions a few species from New Jersey, and in the Curtis Collection are also a few more ; perhaps in all 25 species which might be added to the list.

CHARACEÆ.

(PROVISIONAL LIST.)

COMPILED BY T. F. ALLEN, M. D.

Nitella,

N. flexilis, L.   Canals; common.

N. microcarpa, A. Br.   Morristown.

N. macrocarpa, Allen.   Morris Pond, 1880.

N. gracilis, Sm.   Morris Pond, 1880.

N. tenuissima, Dew.; (forma compacta.)   Morris Pond and Panther Pond, 1880.

N. intermedia, Nordst.   Morris Pond, 1880.

Chara,

C. coronata, Ziz.; *Var.* Schweinitzii.   Common.

C. intermedia, A. Br.   Panther Pond, 1880.

C. Hydropity's, A. Br.; *Var.* septentrionalis, Nordst.   Panther Pond and Morris Pond.

C. sejuncta, A. Br.; (forma elongata.)   Near Panther Pond.

C. sejuncta, A. Br.; (forma condensata.)   Morris Pond.

MARINE ALGÆ.

COMPILED BY A. B. HERVEY.

FLORIDEÆ.

Dasya, Ag.

D. elegans, Ag.

Bostrychia, Mont.

B. rivularis, Harv.

Polysiphonia, Grev.

P. subtilissima, Mont.

P. Olneyi, Harv.

P. Harveyi, Bail.

P. elongata, Grev.

P. violacea, Grev.

P. variegata, Ag.

P. atrorubescens, Grev.

P. fastigiata, Grev.

Rhodomela, Ag.
R. subfusca, Ag.

Chondriopsis, Ag.
C. dasyphylla, Ag.
C. striolata, Ag.
C. tenuissima, Ag.
C. littoralis, Harv.

Grinnellia, Harv.
G. Americana, Harv.

Delesseria, Lam.
D. sinuosa, Lam.

Caloglossa, Harv.
C. Leprieurii, Ag.

Gracilaria, Grev.
G. multipartita.

Corallina, Lam.
C. officinalis, L.

Melobesia, Lam.
M. membranacea, Lam.
M. farinosa, Lam.
M. pustulata, Lam.

Hildenbrandtia, Nardo.
H. rosea, Kütz.

Gelidium, Lam.
G. corneum, Lam.

Hypnea, Lam.
H. musciformis, Lam.

Rhodymenia, Grev.
R. palmata, Grev.

Champia, Ag.
C. parvula, Harv.

Lomentaria, Lyngb.
L. Baileyana, Farlow.

Rhabdonia, Harv.
  R. tenera, Ag.

Polyides, Ag.
  P. rotundus, Ag.

Nemalion, Duby.
  N. multifidum, Ag.

Scinaia, Bivona.
  S. furcellata, Bivona.

Phyllophora, Grev.
  P. Brodiæi, Ag.
  P. membranifolia, Ag.

Gymnogongrus, Mart.
  G. Norvegicus, Ag.

Ahnfeltia, Ag.
  A. plicata, Fr.

Cystoclonium, Kütz.
  C. purpurascens, Kütz.

Chondrus, Lam.
  C. crispus, Lyngb.

Spyridia, Harv.
  S. filamentosa, Harv.

Ceramium, Ag.
  C. rubrum, Ag.
  C. strictum, Harv.
  C. fastigiatum, Harv.

Ptilota, Ag.
  P. plumosa, Ag.
  P. elegans, Bonnem.

Halurus, Kütz.
  H. equisetifolius, Kütz.

Griffithsia, Ag.
  G. Bornetiana, Farlow.

Callithamnion, Lyngb.

    C. tetragonum, Ag.

    C. Baileyi, Harv.

    C. Borreri, Ag.

    C. polyspermum, Ag.

    C. byssoideum, Arn.

    C. Dietziæ, Hooper.

    C. corymbosum, Ag.

    C. versicolor, Ag.

    C. plumula, Lyngb.

    C. Americanum, Harv.

    C. cruciatum, Ag.

    C. Turneri, Ag.

MELANOSPORÆ.

Sargassum, Ag.

    S. vulgare, Ag.

Fucus, L.

    F. nodosus, L. " Rockweed."

    F. vesiculosus, L. " Rockweed."

Laminaria, Lam.

    L. saccarhina, Lam. " Kelp."

Stilophora, Ag.

    S. rhizodes, Ag.

    S. papillosa, Ag.

Striaria, Grev.

    S. attenuata, Grev.

Chorda, Lam.

    C. filum, Stack.

Chordaria, Ag.

    C. flagelliformis, Ag.

    C. divaricata, Ag.

Castagnea, Thuret.

    C. virescens, Thuret.

    C. Zosteræ, Thuret.

Leathesia, Gray.

    L. tuberformis, Gray.

Elachista, Duby.
 E. fucicola, Fr.

Myrionema, Grev.
 M. strangulans, Grev.

Cladostephus, Ag.
 C. spongiosus, Ag.
 C. verticillatus, Ag.

Sphacelaria, Lyngb.
 S. radicans, Ag.
 S. cirrhosa, Ag.

Myriotrichia, Harv.
 M. filiformis, Harv.

Ectocarpus, Lyngb.
 E. firmus, Ag.
 E. siliculosus, Lyngb.
 E. amphibius, Harv.
 E. viridis, Harv.
 E. lutosus, Harv.
 E. Hooperi, Harv.
 E. Dietziæ, Harv.

Dictyosiphon, Grev.
 D. fœniculaceus, Grev.

Desmarestia, Lam.
 D. aculeata, Lam.
 D. viridis, Lam.

Punctaria, Grev.
 P. latifolia.

Phyllitis, Kütz.
 P. fascia, Kütz.

Scytosiphon, Ag.
 S. lomentarius, Ag.

CHLOSPORÆ.

Bryopsis, Lam.
 B. plumosa, Lam.

Enteromorpha, Link.

    E. intestinalis, Link.
    E. compressa, Grev.
    E. clathrata, Grev.

Ulva, Kütz.

    U. latissima, L.
    U. lactuca, L.

Cladophora, Kütz.

    C. rupestris, L.
    C. arcta, Dillw.
    C. glaucescens, Griff.
    C. refracta, Roth.
    C. Morrisiæ, Harv.
    C. albida, Huds.
    C. Rudolphiana, Ag.
    C. gracilis, Griff.
    C. lætivirens, Dillw.
    C. fracta, Fl., Dan.

Chætomorpha, Kütz.

    C. Picquotiana.
    C. ærea, Dillw.

Lyngbya, Ag.

    L. majuscula, Harv.
    L. ferruginea, Ag.
    L. luteo-fusca, Ag.
    L. nigrescens, Harv.

Calothrix, Ag.

    C. confervicola, Ag.
    C. scopulorum, Ag.

Sphærozyga, Ag.

    S. Carmichaelii, Harv.

IN CERTAE SÆDIS.

Porphyra, Ag.

    P. vulgaris, Ag.　" Laver."

Bangia, Lyngb.

    B. fuscopurpurea, Lyngb.

Chantransia, Desv.

    S. virgatula, Thuret.

NOTE.—This Catalogue is not derived from the memoranda of actual collections made upon the New Jersey coast, as such a list ought to be.

So far as I know, no competent botanist has yet explored the waters of that State with the purpose of finding out exactly what marine plants are native to them.

In lack of such data, therefore, I have been obliged to compile this list from printed catalogues of the marine algæ of our coast, assigning to the New Jersey flora only those plants whose known geographical range would naturally bring them within its limits. It is not at all unlikely that a careful survey of these waters would add many species to this list.

FRESH WATER ALGÆ.

COLLECTED BY FRANCIS WOLLE.

FLORIDEÆ.

Lemanea, Bory.

    L. torulosa, Ag.   Attached to stones in swift waters, Bergen Co.

Batrachospermum, Roth.

    B. moniliforme, Roth.   Spring waters, frequent.

    B. vagum, Ag.   Ponds, Burlington Co. and southward.

Chantransia, Fries.

    C. violacea, Kg.   Frequent in shallow streams.

    C. macrospora, Wood.   Abundant in pond, Atsion.

CONFERVOIDEÆ.

Celeochæte, Breb.

    C. scutata, Breb.   Frequent in Lake Hopatcong and other ponds.

    C. soluta, Pringsh.   Lake Hopatcong and other ponds.

    C. orbiculare, Pringsh.   Frequent in ponds.

Œdogonium, Lk.

   Œ. subsetaceum, Kg.   Frequent.

   Œ. paludosum, Wittr.   Perth Amboy.

   Œ. pachydermate, Wittr.   Bound Brook.

   Œ. Wolleanum, Wittr.   Lake Hopatcong, etc.

   Œ. capilliforme, Kg.   Lake Hopatcong.

   Œ. Franklinianum, Wittr.   Franklin Pond.

   Œ. stagnale, Kg.   Bound Brook.

   Œ. ciliatum, Pringsh.   Atsion.

   Œ. polymorphum, Wittr. and Lund.   Bound Brook.

   Œ. sexangulare, Cleve, Lake Hopatcong.

   Œ. platygynum, Wittr.   Atsion.

   Œ. læve, Wittr.   Lake Hopatcong.

   Œ. cryptoporum, Wittr.   Perth Amboy.

   Œ. fonticola, A. Br.   Frequent.

   Œ. capillare, DeC.   Frequent.

   Œ. affine, Rabenh.   Frequent.

   Œ. delicatulum, Kg.   Frequent.

   Œ. echinosporum, A. Br.   Frequent.

Bulbochæte, Ag.

   B. intermedia, de By.   This, and varieties of this species, freely distributed in many ponds.

   B. rectangularis, Wittr.   This, and varieties of this speecies, not rare.

   B. nana, Wittr.   Not infrequent in ponds.

   B. mirabilis, Wittr.   Occasionally met with in ponds.

Cylindrocapsa, Reinsch.

   C. geminella, Wolle.   Not frequent in ponds.

Draparnaldia, Ag.

   D. glomerata, Ag.   Spring waters.

Stigeoclonium, Kg.

   S. tenue, Ag.   Varieties frequent.

Chætophora, Schrank.

   C. pisiformis, Ag.   Not rare in ponds.

   C. endiviæfolia, Ag.   Not rare in ponds.

Aphanochæte, A. Br.

   A. repens, A. Br.   Not rare.

Cladophora, Kg.

    C. fracta, Dillw.   Varieties of this species are frequent.

    C. glomerata, Linn.   Varieties are frequent in running waters.

    C. crispata, Roth.   Frequent in standing waters.

Chroolepus, Ag.

    C. umbrinum, Kg.   On the bark of trees.

    C. aureum, Kg.   On moist rocks.

Ulothrix, Kg.

    U. subtilis, Kg.   Flowing waters.

    U. flaccida, Kg.   Green-houses, etc.

    U. tenuis, Kg.   Rapid waters.

    U. parietina, Kg.   Base of trees.

    U. varia, Kg.   Moist earth.

    U. zonata, (Hormiscia,) Kg.   Streams.

Conferva, Lk.

    C. floccosa, Ag.   Frequent in streamlets.

    C. vulgaris, Rabh.   Trenches.

    C. punctalis, Dillw.   Meadow pools.

    C. bombycina, Ag.   Ponds.

    C. fugacissima, Roth.   Ponds.

Rhizoclonium, Kg.

    R. hieroglyphicum, Ag.   Ponds.

    R. salinum, Kg.   Atlantic City, etc.

    R. fluitans, Kg.   Bound Brook.

    R. major, Wolle.   Perth Amboy.

SIPHONEÆ.

Vaucheria, DC.

    V. sessilis, DC.   Frequent on moist earth.

    V. geminata, DC.   Pools and ponds.

    V. Dillwinii.   Banks of ponds.

    V. Thuretii, Woron.   Soil submerged by tides.

Hydrogastrum, Linn.

    H. granulatum, Desv.   Moist earth, Bergen Co.

PROTOCOCCOIDEÆ.

Volvox, Ehrh.

    V. globator, Ehrh.   Newark, etc.

Pandorina, Bory.

P. Morum, Bory. Not rare.

Hydrodictyon, Roth.

H. utriculatum, Roth. Sluggish waters.

Pediastrum, Mey.

P. tetras, Ehrh.
P. Boryanum, Menegh.
P. pertusum, Kg. Numerous varieties.
P. Ehrenbergii, A. Br.
All of these forms occur frequently in the smaller ponds.

Coelastrum, Naeg.

C. sphæricum, Naeg. Ponds.
C. microporum, Naeg. Ponds.
C. Naegelii, Rabenh. Ponds.

Sorastrum, Kg.

S. spinulosum, Kg. Ponds.

Scenedesmus, Mey.

S. obtusus, Mey. Shallow, stagnant water.
S. acutus, Mey. Shallow, stagnant water.
S. caudatus, Corda. Shallow, stagnant water.

Ophiocytium, Naeg.

O. cochleare, A. Br. Ponds.

Characium, A. Br.

C. subsessilis, Wolle. Cranberry pond.

Protococcus, Ag. Forms of this genus, as far as they have come under my notice, are mere conditions of spores, no true plants— hence omitted.

Polyedrium, Naeg.

P. trigonum, Naeg.
P. aculeatum, Wolle. Not rare in ponds.

Dictyosphærium, Naeg.

D. Ehrenbergianum, Naeg.
D. reniforme, Bulnh.

Tetraspora, Ag.

T. lubrica, Ag. Sluggish streams.
T. gelatinosa, Desv.

Palmella, Lyngb. Forms of this genus I omit. They have no value as perfect plants; they belong to intermediate or arrested life conditions of Algæ.

Porphyridium, Naeg.

P. cruentum, Naeg. Moist earth, green-houses.

Gloeocystis, Naeg. Forms of this genus are not rare, but of doubtful merits as plants.

Rhaphidium, Kg.

R. polymorphum, Fres. Stagnant waters.
R. convolutum, Rabh.
    Var. lunare, Kir.

Nephrocytium, Naeg.

N. Agardhianum, Naeg. Genus and species doubtful.

Pleurococcus, Menegh. Forms of this genus are frequent but of no value; v. note Palmella.

ZYGOSPOREÆ.

Spirogyra, Link.

S. communis, Hass.
S. crassa, Kg.
S. Grevilleana, Hass.
S. inflata, Vauch.
S. intermedia, Rabh.
S. insignis, Hass.
S. longata, Vauch.
S. majuscula, Kg.
S. nitida, Dillw.
S. punctata, Cleve. Atsion.
S. quinina, Ag.
S. rivularis, Hass.
S. stagnalis, Hilse.
S. varians, Hass.
S. Weberi, Kg.
S. fluviatilis, Hilse.

These forms and the following appear to be quite generally distributed through the State, in streams and ponds.

Zygnema, Ag.

Z. cruciatum, Ag. Frequent.
Z. insigne, Kg. Frequent.

Z. stellinum, Ag.   Frequent.
Z. tenue, Kg.   Frequent.
Z. Vaucheria, Ag.   Frequent.

Zygogonium, Kg.
Z. Agardhii, Rabh.   Not rare.
Z. pectinatum, Kg.   Not rare.

Mougeotia, De By.
M. lævis, Archer.   Franklin.

Mesocarpus, Hass.
M. scalaris, Hass.   Frequent.
M. nummuloides, Hass.   Frequent.

Craterospermum, A. Br.
C. lætevirens, A. Br.   Green Pond, etc.

Staurospermum, Kg.
S. Capucinum, Kg.   Pleasant Mills, Atsion, etc.

Hyalotheca, Ehrb.
H. mucosa, Ehrb.   Frequent.
H. dissiliens, Breb.   Frequent.

Bambusina, Kg.
B. Brebissonii, Kg.   Frequent.

Desmidium, Ag.
D. Swartzii, Ag.   Frequent.
D. Aptogonium, Breb.

Aptogonum, Ralfs.
A. Baileyi, Ralfs.   Pleasant Mills, Ocean Co., etc.

Sphærozosma, Corda.
S. vertebratum, Ralfs.   Frequent in ponds.
S. excavatum, Ralfs.   Frequent in ponds.
S. filiforme, Ehrb.   Frequent in ponds.
S. pulchellum, Archer.   Frequent in ponds.
S. pulchrum, Bail.   Frequent in ponds.
S. secedens, De By.   Frequent in ponds.
S. serratum, Bail.   Frequent in ponds.

Mesotænium, Naeg.
M. micrococcum, Kg.   Moist earth.

Penium, Breb.

    P. digitus, Breb. Frequent.
    P. margaritaceum, Breb. Frequent.
    P. interruptum, Breb. Frequent.
    P. Closteroides, Ralfs. Frequent.
    P. polymorphum, Perty. Frequent.
    P. Brebissonii, Ralfs. Frequent.

Closterium, Nitzsch.

    C. juncidum, Ralfs.
    C. lunula, Ehrb.
    C. acerosum, Ehrb.
    C. turgidum, Ehrb.
    C. striolatum, Ehrb.
    C. costatum, Corda.
    C. lineatum, Ehrb.
    C. decorum, Breb.
    C. Dianæ, Ehrb.
    C. Jenneri, Ralfs.
    C. Venus, Kg.
    C. Ehrenbergii, Menegh.
    C. Leibleinii, Kg.
    C. rostratum, Ehrb.
    C. setaceum, Ehrb.
    C. Ralfsii, Breb.

All these liberally distributed in ponds throughout the State.

Calocylindrus, D. By.

    C. Ralfsii, Kg. Frequent.
    C. palangula, Breb. Frequent.
    C. cucurbita, Breb. Frequent.
    C. curtus, Breb. Frequent.
    C. connatus, Breb. Frequent.

Docidium, Breb.

    D. Baculum, Breb. Frequent.
    D. constrictum, Bail. Frequent.
    D. gracile, Bail. Frequent.
      Var. bidentatum, Nordt.
    D. verticillatum, Bail. Frequent.
    D. dilatatum, Lund. Frequent.
    D. spinosum, Wolle. Dennisville.
    D. undulatum, Bail. Dennisville.
    D. nodosum, Bail. Dennisville.

Pleurotænium, Naeg.

  P. crenulatum, Ehrb. Frequent.
  P. truncatum, Breb. Frequent.
  P. Trabecula, Naeg. Frequent.
  P. clavatum, Kg. Frequent.
  P. coronatum, Breb. Frequent.

Tetmemorus, Ralfs.

  T. Brebissonii, Ralfs. Frequent.
    Var. turgidus, Ralfs.
  T. granulatus, Ralfs. Frequent.
  T. laevis, Ralfs. Frequent.
  T. giganteus, Wood. Atsion.

Cosmarium, Corda.

  C. amœnum, Breb. Hammonton.
  C. bioculatum, Breb. Common.
  C. Botrytis, Menegh. Common.
  C. Biretrum, Breb. Bergen Co.
  C. Brebissonii, Menegh. Atlantic Co.
  C. cælatum, Ralfs. Frequent.
  C. conspersum, Ralfs. Frequent.
  C. crenatum, Ralfs. Frequent.
  C. cucumis, Corda. Frequent.
  C. cylindricum, Ralfs.
  C. dentatum, Wolle. Dennisville.
  C. granatum, Breb. Frequent.
  C. Hammeri, Renisch. Frequent.
  C. irregularis, Wolle. Budd's Lake.
  C. margaritiferum, Menegh. Frequent.
  C. Meneghinii, Breb. Frequent.
    Var. nanum, Wille.
  C. moniliforme, Ralfs. Frequent.
  C. margaritum, Wolle. Dennisville.
  C. orbiculatum, Ralfs. Frequent.
  C. ornatum, Ralfs. Frequent.
  C. ovale, Ralfs. Frequent.
  C. phaseolus, Breb. Frequent.
  C. Porterianum, Archer. Frequent.
  C. pyramidatum, Breb. Frequent.
  C. sportella, Breb. Lake Hopatcong.
  C. smolandicum, Lund. Frequent.
  C. sublobatum, Archer. Frequent.
  C. sexangulare, Lund. Split Rock Pond.

C. tumidum, Lund.   Split Rock Pond.
C. tetrophthalmum, Kg.   Frequent.
C. Turpinii, Breb.   Not rare.

Xanthidium, Ehrb.

X. aculeatum, Ehrb.   Lake Hopatcong, etc.
X. armatum, Breb.   Hammonton, Pleasant Mills, etc.
X. antilopaeum, Kg.   Frequent.
X. cristatum, Breb.   Hammonton.
X. fasciculatum, Ehrb.   Frequent.

Arthrodesmus, Ehrb.

A. convergeus, Ehrb.   Frequent.
A. fragilis, Wolle.   Hammonton, etc.
A. incus, Hass.   Hammonton, etc.
A. octocornus, Ehrb.   Hammonton, etc.

Euastrum, Ehrb.

E. ampullaceum, Ralfs.   Frequent.
E. attenuatum, Wolle.   Budd's Lake.
E. abruptum.
   Var. evolutum, Nordt.   Tom's River.
E. affine, Ralfs.   Frequent.
E. binale, Ralfs.   Frequent.
   Var. insulare, Wittr.   Frequent.
E. crassum, Breb.   Frequent.
E. circulare, Hass.   Frequent.
E. Didelta, Turp.   Frequent.
E. elegans, Breb.   Common.
E. formosum, Wolle.   Tom's River.
E. gemmatum, Breb.   Split Rock Pond.
E. humerosum, Ralfs.   Frequent.
E. insigne, Hass.   Frequent.
E. intermedium, Cleve.   Dennisville.
E. oblongum, Ralfs.   Not rare.
E. pectinatum, Breb.   Not rare.
E. Ralfsii, Rabh.   Not rare.
E. rostratum, Ralfs.   Not rare.
E. pinnatum, Ralfs.   Not rare.
E. spinosum, Ralfs.   Not rare.

Micrasterias, Ag.

M. arcuata, Bail.   Pleasant Mills, etc.
M. Americana, Ralfs.   Frequent.
   Var. recta, Wolle.   Dennisville.

M. Baileyi, Ralfs.   Princeton, etc.
M. Crux Melitensis.
   Var. Ehrb.   Dennisville.
M. crenata, Breb.
M. denticulata, Breb.   Frequent.
M. disputata, Wood.   Frequent.
M. furcata, Ag.   Frequent.
M. fimbriata, Ralfs.   Frequent.
   Var. apiculata, Menegh.
M. Jenneri, Ralfs.   Occasional.
M. Kitchelii, Wolle.   Dennisville.
M. muricata, Ralfs.   Rather rare.
M. mucronata, Dixon.   Brown's Mills.
M. oscitans, Ralfs.   Frequent.
M. pinnatifida, Kg.   Frequent.
M. pseudofurcata, Wolle.   Split Rock Pond, etc.
M. rotata, Ralfs.   Frequent.
M. radiosa, Ag.-Sol., Ehrb.   Frequent.
M. ringens, Bail.   Frequent.
M. truncata, Corda.   Frequent.
M. Torreyi, Bail.   Split Rock Pond.

Staurastrum, Mey.
   S. arcuatum, Nosdt.   Split Rock Pond.
   S. aculeatum, Ehrb.   Frequent.
   S. alternans, Breb.   Frequent.
   S. arachne, Ralfs.   Frequent.
   S. aristiferum, Ralfs.   Hammonton, etc.
   S. asperum, Breb.   Not rare.
   S. Avicula, Breb.   Frequent.
   S. bifidum, Ehrb.   Frequent.
   S. brachycerum, Bieb.   Frequent.
   S. brachiatum, Ralfs.   Atsion.
   S. brevispinum, Breb.   Atsion, etc.
   S. cyrtocerum, Breb.   Split Rock Pond.
   S. cuspidatum, Breb.   Franklin.
   S. Dickiei, Ralfs.   Frequent.
   S. dejectum, Breb.   Frequent.
   S. echinatum, Breb.   Frequent.
   S. eustephanum, Ralfs.   Split Rock Pond.
   S. furcegerum, Breb.   Split Rock Pond.
   S. gracile, Ralfs.   Frequent.
   S. geminatum, Nordt.   Split Rock Pond.
   S. hirsutum, Ehrb.   Frequent.
   S. Helencanum, Wolle.   Split Rock Pond.

S. Haabœliense, Wille.   Split Rock Pond.
S. leptocladon, Nordt.   Cranberry Pond, etc.
S. margaritaceum, Ehrb.   Frequent.
S. macrocerum, Wolle.   Atsion, etc.
S. muricatum, Breb.   Tom's River.
S. munitum, Wood; Arctiscon, Ehrb.
S. Novæ-Cæsareæ, Wolle.   Hammonton.
S. ophiura, Lund.   Split Rock Pond.
S. orbiculare, Ehrb.
S. odontodum, Wolle.   Split Rock Pond.
S. pentacladum, Wolle.   Split Rock Pond.
S. paradoxum, Mey.   Frequent.
S. polymorphum, Breb.   Frequent.
S. polytrichum, Perty.   Frequent.
S. pulchrum, Wolle.   Brown's Mills, Split Rock Pond.
S. pusillum, Wolle.   Brown's Mills.
S. punctulatum, Breb.   Frequent.
S. pygmæum, Breb.   Frequent.
S. rugulosum, Breb.   Tom's River.
S. scabrum, Breb.   Tom's River.
S. senarium, Ehrb.   Rare.
S. setigerum, Cleve.   Frequent.
S. saxonicum, Bulnh.   Budd's Lake.
S. Sebaldi, Reinsch.   Occasional.
S. teliferum, Ralfs.   Frequent.
S. tricornutum, Wolle.   Hammonton.
S. tricorne, Menegh.   Frequent.
S. terebrans, Nordt.   Atsion.
S. vestitum, Ralfs.   Frequent.
S. trifidum, Nordt.   Atsion, etc.

SCHIZOSPOREÆ.

Calothrix, Ag., Thur.

C. Orsiniana.   Morris Pond, etc.
C. radiosa.
    Var. fuscescens, Kg.   Green Pond.
C. Meneghiniana, Kg.   Atsion, etc.
C. lacunosa, Wolle.   Split Rock Pond.

Mastigonema, Fischer.

M. ærugineum, Kir.   Frequent.
M. cæspitosum, Kg.   Bergen Co.
M. velutinum, Wolle.   Perth Amboy.
M. pluviale, A. Br.   Ponds.

Gloiotrichia, (Rivularia,) Ag.

    G. pisum, Ag. Ponds.

    G. natans, Thur. Ponds.

Rivularia, Roth.

    R. radians, Thur. Frequent in ponds.

Isactis, Thur.

    I. fluviatilis, Kg. Green Pond.

Scytonema, Ag.

    S. Austinii, Wood. Little Falls.

    S. calotrichoides, Kg.

      Var. natans, Rabh. Brown's Mills, etc.

    S. cinereum, Menegh. Goodwinsville.

    S. gracile, Kg. Morris Pond, etc.

    S. myochrous, Ag. Closter.

    S. Naegeli, Kg. Goodwinsville and Closter.

    S. natans, Breb. Hammonton.

    S. truncicola, Rabh. Bergen Co.

    S. tolypotrichoides, Kg. Wet rocks. Not unfrequent.

Symphyosiphon, Kg.

    S. Hofmanni, Kg. Moist earth and rocks.

    S. tenuis, Kg. Palisades.

Tolypothrix, Kg.

    T. ægagropila, Kg. Frequent.

    T. distorta, Kg. Ponds.

    T. bombycina, Wolle. Growing on rocks, Lake Hopatcong.

Sirosiphon, Kg.

    S. ocellatus, Kg. Frequent in swampy places.

    S. compactus, Ag., Kg. Moist rocks.

    S. pulvinatus, Breb. Moist rocks.

    S. coralloides, Kg. Green Pond, etc.

Hapalosiphon, Naeg.

    H. Braunii, Naeg. Atsion, etc.

    H. fuscescens, Kg. Frequent in ponds.

    H. Brebissonii, Kg. Moist earth.

    H. tennissimus, Grun. Ponds.

Nostoc, Vauch.

    N. spheroides, Kg.

    N. rupestre, Kg.

N. cæruleum, Lyng.
N. commune, Vauch.
N. cristatum, Bail-Alpinum, Kg.
N. comminutum, Kg.

These and other varieties are not infrequent.

In the Bulletin of the Torrey Botanical Club, April, 1879, I noted, "A Nostoc the matrix of Scytonema." The probability is that many of the recorded forms are mere varying phases of the same, and that all the Nostocs are undeveloped forms, representatives of arrested or intermediate life conditions of higher stages of development.

Anabæna, Bory.

A. flosaqua, Kg.
A. circinalis, Rabh.   Probably a variety of the former; frequent; very abundant at Dennisville.

Trichormus, Allman.

T. incurvus, Allman.   On bark of old logs in swamp near Closter.

Sphærozyga, Ag.

S. polysperma, Rabh.   Bound Brook, etc.
S. saccata, Wolle.   Cranberry Pond.

Cylindrospermum, Kg.

C. macrosporum, Kg.   Wet places, old wood, etc.

Lyngbya, Ag.

L. æruginosa.   Perth Amboy, etc.
L. Æstuarii, Jurg.   Near sea coast.
L. Wollei, Farlow.   Lake Hopatcong, Swartwout Pond.
L. obscura.   Ponds.

Symploca, Kg.

S. lucifuga, Harv.   Bergen Co.

Microcoleus, Desm.

M. terrestris, Desm.=Chthonoblastus Vaucheri, Kg.=Ch. repens, Kg.   Moist earth.

Inactis, Kg.

I. Austinii, Wolle.   Little Falls.

Oscillaria, Bosc.

O. brevis, Kg.   Marshes.
O. Frœlichii, Kg.   Cape May Co., etc.

O. gracillima, Kg. Small ponds.
O. limosa, Ag. Wet earth.
O. nigra, Vauch. Frequent.
O. natans, Kg. Panther Pond, etc.
O. princeps, Vauch. Dennisville.
O. rupestris, Ag. Palisades, banks of Delaware.
O. tenuis, Ag. Stagnant waters.

Phormidium, Kg.

P. cataractum, Rabh. Rapid waters.
P. Julianum, Menegh.
P. Lyngbyaceum, Kg. Bergen Co.
P. Naveanum, Grun. Bergen Co.
P. rufescens, Ag. Bound Brook.
P. vulgare, Kg. Frequent.
  Var. publican, Kg. Frequent.

Leptothrix, Kg.

L. ochracea, Kg. Ditches.
L. æruginea (Hypheothrix), Rabh. Frequent.

Glœothece, Naeg.

Aphanothece, Naeg.

A number of varieties have been found, but as the genera are of doubtful merit they are not enumerated.

Merismopedia, Mey.

M. convoluta, Breb. Frequent in ponds.

Microcystis, Kg.

Polycystis, Kg.

Gloeocapsa, Naeg.

Aphanocapsa, Naeg.

Chroococcus, Naeg.

These genera are represented by many varieties, but all are of doubtful value as plants. They represent spore conditions of higher forms, and therefore are not enumerated.

INDEX OF GENERA.

Podostemon	83		Samolus	61
Pogonatum	147		Sanguinaria	7
Pogonia	95	Racomitrium	144	Sanicula	38
Polanisia	11, 128	Radula	157	Saponaria	14
Polemonium	65, 127	Radulum	173	Sarcoscyphus	154
Polyactis	188	Ramalina	159	Sargassum	207
Polycarpon	128	Ranunculus	2, 128	Sarracenia	6
Polycystis	223	Raphanus	11	Sasbania	130
Polyedrium	213	Rapistrum	128	Sassafras	84
Polygala	13	Reseda	11, 128	Saururus	84
Polygonatum	102	Reticularia	177	Saxifraga	33
Polygonella	83	Rhabdonia	206	Scandix	130
Polygonum	82, 127	Rhabdoweissia	142	Scapania	154
Polyides	206	Rhaphidium	214	Scenedesmus	213
Polymnia	49, 126	Rhamnus	22	Scheuchzeria	98
Polypodium	135	Rhexia	35	Schizæa	137
Polypogon	117	Rhinotrichum	187	Schizophyllum	170
Polyporus	171	Rhizoclonium	212	Schollera	107
Polypremum	64	Rhizopogon	176	Schwalbea	73
Polysaccum	177	Rhododendron	60	Scinaia	106
Polysiphonia	204	Rhodomela	205	Scirpus	110, 133
Polythrincium	187	Rhodora	60	Scleria	112
Polytrichum	147	Rhodymenia	205	Scleroderma	176
Pontederia	106	Rhus	23	Sclerolepis	44
Populus	90	Rhynchosia	130	Scleranthus	17
Porothelium	172	Rhyncospora	111	Scolymus	131
Porphyra	209	Rhyncostegium	151	Scoparia	132
Porphyridium	214	Rhytidium	154	Scrophularia	70, 132
Portulaca	17, 129	Rhytisma	195	Scutellaria	78
Potamogeton	99	Ribes	32	Scytonema	221
Potentilla	29	Riccia	158	Scytosiphon	208
Poterium	29, 130	Richardsonia	130	Sedum	33
Pottia	143	Ricinus	133	Selaginella	139
Preissia	158	Rinodina	163	Seligeria	143
Propolis	194	Rivularia	221	Selinum	38
Proserpinaca	34	Robinia	24	Sendtnera	156
Protococcus	213	Rœstelia	184	Senebiera	11
Prunus	28	Rosa	31	Senecio	54, 131
Psalliota	169	Roubieva	133	Sepedonium	189
Pterigynandrum	149	Rubus	30	Septoria	181
Pteris	135	Rudbeckia	50	Septonema	182
Ptilota	206	Rumex	83, 133	Septosporium	187
Ptychomitrium	145	Ruppia	99	Sericocarpus	46
Puccinia	183	Russula	170	Sesamum	132
Punctoria	208			Sesuvium	37
Pycnanthemum	76			Setaria	125, 134
Pylaisia	150	**S**		Sherardia	131
Pyrenula	167			Sicyos	37
Pyrola	60	Sabbatia	64	Sida	19, 129
Pyxidanthera	61	Sagedia	167	Silene	15, 129
Pyxine	161	Sagina	16	Sirosiphon	221
		Sagittaria	98	Sisymbrium	9, 128
		Salicornia	81	Sisyrinchium	97
		Salix	90, 127, 133	Sium	40
Q		Salsola	81	Smilacina	102
		Salvia	77, 126	Smilax	105
Quercus	88	Sambucus	42	Solanum	69, 132

www.ingramcontent.com/pod-product-compliance
Lightning Source LLC
Chambersburg PA
CBHW021400210326

41599CB00011B/954